CARBON DIOXIDE FLOODING

BASIC MECHANISMS AND PROJECT DESIGN

CARBON DIOXIDE FLOODING

BASIC MECHANISMS AND PROJECT DESIGN

MARK A. KLINS
The Pennsylvania State University

International Human Resources Development Corporation • Boston

Library of Congress Cataloging in Publication Data

Klins, Mark A., 1953-
  Carbon dioxide flooding.

  Bibliography: p.
  Includes index.
  1. Oil field flooding.   2. Carbon dioxide.   I. Title.
TN871.K5554   1984   622'.3382   83-26494
ISBN 0-934634-44-0

Printed in the United States of America

CONTENTS

ACKNOWLEDGMENTS

The author wishes to thank a number of his colleagues who have made this text possible. Special appreciation is given to Dr. Michael B. Moranville, Dr. Thomas L. Gould, Charles W. Bloomquist, and G. Thomas Smart. Their work in helping prepare the original text notes for Intercomp, Inc. was invaluable. Tribute is also paid to Dr. Turgay Ertekin for his aid in laying the groundwork of the reservoir modeling chapter.

Recognition is also extended to Intercomp, Inc. for their financial assistance during this project. Lastly, special thanks is given to Dr. R. Paul Hackleman and Dr. S. M. Farouq Ali. As friends and counsels they have made this book a reality.

Mark A. Klins

INTRODUCTION AND OVERVIEW

Many techniques have been investigated in the laboratory and the field for improving oil recovery. Today's technology and economics leave two barrels of oil in the ground for each barrel recovered. To understand the import of this statement, consider that an increase of only 1% in oil recovery efficiency for the United States represents about 4 billion barrels—more than it produces in one year.

To tap these reserves, common techniques include steam injection, in situ combustion, chemical flooding, and caustic injection (chapter 1). Currently, however, due to its wide applicability, there is a great deal of interest in carbon dioxide (CO_2) for the recovery of both light and heavy oils.

This text first discusses why CO_2 mobilizes and displaces oil. Several mechanisms play a role in increasing oil recovery. These mechanisms, discussed in chapter 2, are dependent on reservoir temperature, displacement pressure, and the properties of the crude oil. Besides augmenting natural drive mechanisms by the injection of energy, CO_2 lowers oil viscosity, adds gas into solution for oil swelling and additional gas drive, as well as creating potential miscibility.

Coupled with the discussion of why oil is displaced by CO_2 must be an analysis of CO_2-water-crude oil interaction. Chapter 3 examines this topic and points out not only how CO_2 reacts with formation fluids but also the type of information that is necessary to predict accurately a reservoir's response to CO_2 injection. This chapter also discusses the types of laboratory tests that are needed to obtain this information.

Core, log, production history, and pressure-volume-temperature (PVT) data are then input to predict the response of a CO_2 flood. Several prediction tools are available to the engineer. All cases involve reservoir simulation, but no reliable analytical techniques have been designed to date. The prediction methods most commonly applied are discussed in chapter 4 and include:

black oil simulation for very low pressure displacements,
miscible models for very high CO_2 injection pressures, and
compositional modeling for the vast majority of intermediate pressure applications.

Chapter 5 overviews the types of reservoirs in which CO_2 injection has been successful. Included are screening guides of key reservoir parameters, why these variables serve as a limitation on the applicability of CO_2, and how these key parameters can be determined.

Chapter 6 describes the field design of a CO_2 injection project. Reservoirs must be screened and modeled, pilot tests implemented, and field equipment designed. Interwoven into this chapter and the previous chapters are field examples of data gathering, modeling techniques, and economic analysis.

There is little doubt that enhanced oil recovery will make significant contributions to future oil recovery, and CO_2 injection will play a major role in augmenting that recovery. Under this hypothesis, we have created this text.

1 REVIEW OF ENHANCED OIL RECOVERY

A new awareness of energy needs is forcing a reassessment of how the world will fuel the economy of its future generations. For over 150 years we have, with increasing regularity, been withdrawing cheap, abundant fuel from earth's finite energy bank.

Today, petroleum is produced worldwide at a rate of 30,000 gallons per second, 9,000 of which is used in the United States every second of every day in the year. While encompassing only 6% of the world's population, Americans use 25% of all the world's energy—30% of its oil. This oil plays a major role in the U.S. energy spectrum and provides almost half of the U.S. energy diet, as shown in figure 1.1. Not only does oil carry a leadership role in today's world, but also it is expected to maintain that stature for the coming decades ("Winter's Legacy," 1977).

Although oil prices have increased roughly tenfold in the last decade, figure 1.2 shows an ever-growing demand for petroleum products. Coupled with this demand for crude oil is the inability to produce adequate stocks of petroleum domestically. These two facts identify a shortfall made up by imports of nearly 2 billion barrels per year. The three major reasons for industry's inability to meet this growing demand are

1. price,
2. setbacks in discovering new oil, and
3. the inefficiency with which we remove oil from the ground.

The price of oil in the United States has been regulated in the past to an artificially low level when compared to the world oil market. While this has aided the U.S. consumer and kept demand high, it has made continued production from many fields impossible. Not only is it becoming increasingly costly to develop and produce existing reserves, but also the task of discovering new reserves is much more difficult now than in the past.

Figure 1.3 relates estimates of ultimate recovery of crude oil from reservoirs to their years of discovery. Except for the 10 billion barrels at Prudhoe Bay, the discovery rate in the United States has averaged fewer than 1 billion barrels a year for the past twenty years. Figure 1.3 also shows our domestic production of crude oil. Note that in recent

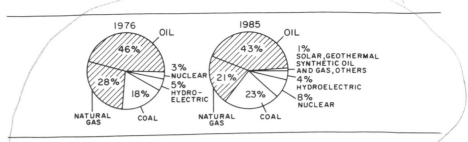

Figure 1.1 Share of all energy used in the United States ("Winter's Legacy" 1977).

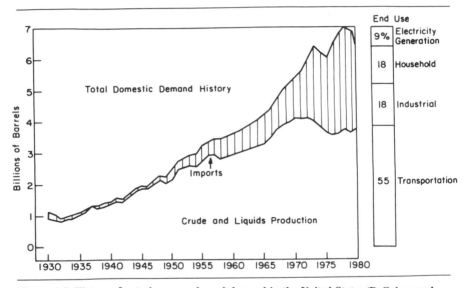

Figure 1.2 History of petroleum supply and demand in the United States (DeGolyer and MacNaughton 1981).

years, U.S. oil production has hovered near 3 billion barrels a year, signaling an alarming disparity between U.S. consumption and the discovery of new reserves.

Coupled with the early discovery of most of our large oil fields was our ability to recover a significant portion of the oil in them. Many of these early reservoirs had high permeability and strong natural water drives that made estimates of future recovery from newly discovered oil fields optimistic. Figure 1.4 illustrates recovery efficiency through time. Note that the trend from 1920 through 1940 clearly indicates future recovery efficiencies to be over 50% of the oil in place. Unfortunately, newer fields were naturally less efficient than their predecessors. Figure 1.4 also shows that we recover only 30% of the oil we discover; that is, for every barrel of oil produced, two barrels remain in the ground.

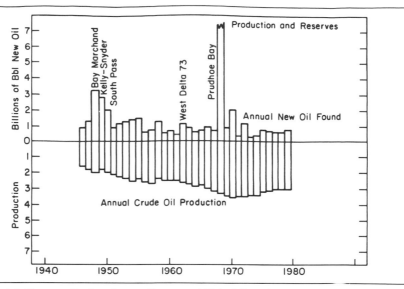

Figure 1.3 The contribution of new oil to proved United States' crude reserves (Moody 1978).

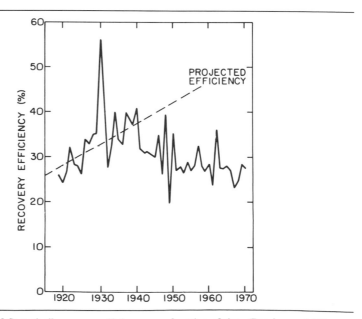

Figure 1.4 United States' oil recovery efficiency as a function of time (Doscher and Wise 1976).

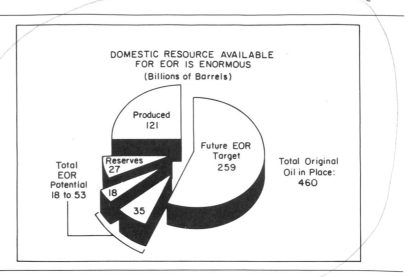

Figure 1.5 Tertiary recovery stakes in the United States (Johnson 1982).

1.1   **REMAINING RESERVES IN THE UNITED STATES.** To revise this scenario of increasing demand and price tied with decreasing discoveries and production, industry must tap these large reserves of oil that lie trapped in reservoirs discovered years ago. What once was considered unrecoverable oil should prove to be technically and economically attractive in the near future.

Of the 460 billion barrels of oil discovered in the United States since 1859, present production techniques should result in a recovery of only 148 billion barrels (about 32%). Conversely, when our present fields are abandoned, some 312 billion barrels of oil—68%—will remain locked in the ground, as shown in figure 1.5.

1.2   **POTENTIAL PRODUCTION OF ENHANCED OIL RECOVERY.** As shown in figure 1.5, it is estimated that from 18 to 53 billion barrels of oil may be produced by various enhanced oil recovery (EOR) techniques. Considering that the United States has 27 billion barrels of reserves labeled recoverable by conventional techniques, the additional amount made possible through enhanced recovery techniques is significant. Table 1.1 presents a breakdown of these potential reserves by recovery type. CO_2 flooding shows the widest applicability with predicted production up to 21 billion barrels. Also listed are estimates of the relative costs of each process and its oil recovery efficiency.

Despite the fact that laboratory and pilot tests have demonstrated the ability of the various techniques to produce additional oil, figures 1.6 and 1.7 show that only 4% of the nation's production is presently derived from EOR, and two-thirds of this is attributed to one process—steam recovery of heavy oils in California. In order for an EOR

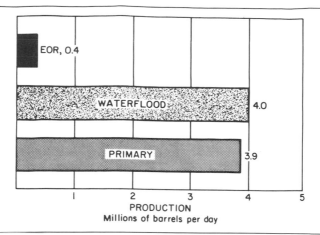

Figure 1.6 Current crude oil production (1980) (Johnson 1982).

Table 1.1 Potential Reserves and Costs Listed by Enhanced Recovery Process

Recovery Process	Cost Per Barrel (dollars)	Recovery Efficiency (percent)	Potential (billions of barrels)	
			Current Technology	Advanced Technology
Steam drive	21–35	25–64	4.0	13.0
In situ combustion	25–36	28–39	1.6	0.4
Polymers	22–46	0–4	0.3	0.6
Micellar/Polymer	35–46	30–40	5.0	7.0
CO_2 Miscible	26–39	15–19	7.0	14.0

Source: H. R. Johnson, *Outlook for Enhanced Oil Recovery,* DOE internal report, Bartlesville, Okla. (1982).

process to be viable, it not only must increase oil recovery but also must do it in an economically attractive manner.

Economic attractiveness implies not only that the cost of implementing an EOR process be less than the value of resulting increased oil recovery but also that the timing of the expenditures and revenues be such that an acceptable present value profit can be made. Currently, it is more a question of the profitability rather than the technical feasibility of most EOR processes that is restricting their application.

As table 1.2 shows, interest in EOR projects has increased over the past few years, suggesting that the economic climate may be becoming more favorable to such processes. However, lengthy pilot tests, higher oil prices, new tax incentives for enhanced oil producers, and stabilized government regulations are required before EOR can become a viable energy alternative. Where enhanced oil accounts for 385,000 barrels per day today, it may well contribute, as shown in figure 1.8, between 1 and 4 million

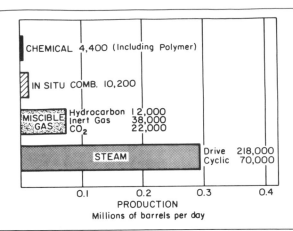

Figure 1.7 EOR production by process (1982) ("Annual Production Report" 1982).

Table 1.2 Number of Active Domestic EOR Projects

Method	1971	1974	1976	1978	1980	1982
Thermal Methods						
Steam	53	64	85	99	133	118
In situ combustion	38	19	21	16	17	21
Chemical Methods						
Micellar-polymer	5	7	13	22	14	20
Polymer	14	9	14	21	22	47
Caustic	0	2	1	3	6	10
Gas Methods						
CO_2 miscible	1	6	9	14	17	28
Hydrocarbon miscible	21	12	15	15	9	12
Other gases	0	1	1	6	8	10

Source: "Annual Production Report," *Oil and Gas J.* (Apr. 5, 1982).

barrels per day in 1995. For perspective, while today's enhanced recovery rate makes up 4% of domestic production, the upper level 1995 rate is equal to nearly half of the current total U.S. domestic crude oil production.

1.3   FACTORS DETERMINING RECOVERY EFFICIENCY. A simplistic model for estimating overall recovery involves factoring the recovery efficiency into individual process efficiencies

$$E_R = E_A \cdot E_V \cdot E_D \cdot E_M, \tag{1.1}$$

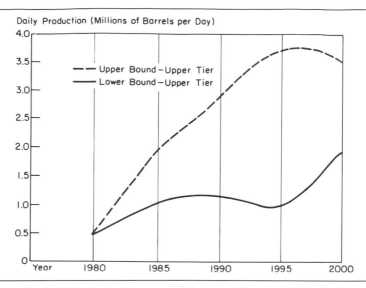

Figure 1.8 Potential EOR production in the United States (*Potential and Economics* 1976).

where

E_R = overall recovery efficiency,
E_A = areal sweep efficiency,
E_V = vertical sweep efficiency,
E_D = displacement efficiency,
E_M = mobilization efficiency.

A detailed discussion of the various individual efficiencies and methods for estimating them is deferred to chapter 2. Here, we wish to define the recovery factor and indicate how various EOR techniques attempt to increase or accelerate oil recovery by modifying one (or more) of these individual efficiencies.

The individual terms in equation (1.1) are defined as follows:

Areal sweep efficiency (E_A) is the fractional area of the field that is invaded by an injected fluid. The major factors determining areal sweep are fluid mobilities (permeability/viscosity), pattern type, areal heterogeneity, extent of field development, and total volume of fluid injected. Figure 1.9 illustrates the effect of mobility ratio on sweep efficiency for a five-spot pattern. Note that sweep efficiency increases from 0 at the start of fluid injection to 1 at essentially infinite pore volumes injected. Of course, some economic limit will be reached before 100% sweep of the reservoir is achieved.

Vertical sweep efficiency (E_V) is the fraction of the vertical section that is contacted by injected fluids and is primarily a function of the vertical heterogeneity and the degree of gravity segregation (dependent on dip and horizontal and vertical permeability) in the reservoir. Figure 1.10 illustrates the effect of these factors on vertical sweep. Like areal sweep, vertical sweep will increase with increasing volumes of total injection, starting at 0 and having a maximum value of 1.

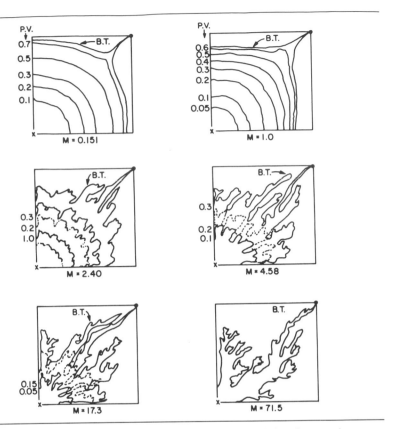

Figure 1.9 Areal sweep efficiency for different mobility ratios and injected pore volumes until breakthrough (Habermann 1960).

Mobilization efficiency (E_M) is defined as the fraction of the oil in place at the start of a recovery process that ultimately could be recovered by that process and is given by:

$$E_M = \frac{S_{oi}/B_{oi} - S_{orp}/B_{of}}{S_{oi}/B_{oi}}, \tag{1.2}$$

where

S_{oi} = oil saturation at start of project,
B_{oi} = oil formation volume factor at start of project,
S_{orp} = residual oil to process,
B_{of} = oil formation volume factor at end of process.

Thus, for a waterflood, S_{oi} would be the oil saturation at the start of water injection, and S_{orp} would be the residual oil after waterflooding (S_{orw}) as measured in a flood pot.

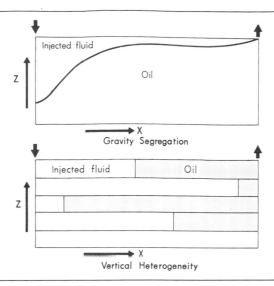

Figure 1.10 Examples of vertical sweep efficiency when gravity segregation or heterogeneity are present.

For a tertiary recovery process, S_{oi} would be waterflood residual (S_{orw}), and S_{orp} would be the residual to that specific process.

Unlike the other individual efficiencies, E_M is taken to be independent of cumulative injection and can be viewed as the maximum oil recovery that could be technically (as opposed to economically) mobilized. The factors that determine the mobilization efficiency $(S_{orp}$ and $B_{of})$ will depend on the particular process but include the ratio of capillary to viscous forces and the phase behavior of the crude oil and injected fluid.

Displacement efficiency (E_D) is the fraction of the mobile oil in the swept zone that has been displaced and is a function of the volume injected, the fluid viscosities, and the relative permeability curves of the rock. Figure 1.11 illustrates the effect of viscosity on displacement efficiency. Like areal and vertical sweep efficiency, displacement efficiency (as defined here) varies from 0 at the start of injection to 1 at infinite throughput.

1.4   ENHANCED OIL RECOVERY METHODS. The various processes referred to as EOR methods seek to improve one or more of the individual efficiencies in equation (1.1). Here, "improve" means not only to increase the amount of recovery but also to accelerate that recovery. In order for a specific process to be viable, it not only must improve recovery but also must do it in an economically attractive manner.

Table 1.3 indicates which of the individual efficiencies the various EOR techniques are designed to improve. In this table, P indicates the major effects of a particular process and S indicates a significant secondary effect. For example, the primary effect of steam flooding is to improve displacement efficiency by reducing the oil viscosity.

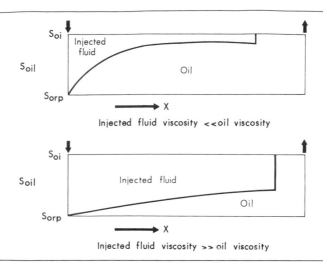

Figure 1.11 Effect of viscosity on displacement efficiency.

Table 1.3 Recovery Efficiency Terms Affected by Various EOR Processes

	Areal Sweep	Vertical Sweep	Displacement Efficiency	Mobilization Efficiency
Miscible recovery				P[a]
Polymer recovery	S[b]	S	P	
Micellar-polymer	S	S	S	P
Caustic				P
Steam			P	S
Combustion			P	S
Infill drilling	P	P		
CO$_2$ flooding				
Miscible				P
Immiscible			P	S

[a]P = primary effect.
[b]S = significant secondary effect.

However, the residual oil to steam is frequently significantly lower than the residual oil to water, and hence, a secondary effect of steam flooding is to increase the mobilization efficiency.

A brief introduction to each of these recovery techniques is now in order, including a description of the process, advantages, and disadvantages. For further information, one can refer to a series of articles presented by Herbeck et al. in *Petroleum Engineer* (1976–1977).

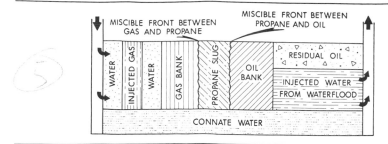

Figure 1.12 LPG Slug Process (Herbeck et al. 1976). Reprinted by permission of
PETROLEUM ENGINEER International.

1.4.1   Hydrocarbon Miscible Processes

LIQUEFIED PETROLEUM GAS (LPG) MISCIBLE SLUG. Miscible slug displacement usually refers to the injection of some liquid solvent that is miscible upon first contact with
the resident crude oil. Specifically, the process uses a slug of propane or other LPG (2
to 5% PV [pore volume]) tailed by natural gas, inert gas, and/or water. The solvent, if
designed correctly, should bank oil and water ahead of it and completely displace all
contacted oil.

Although lower injection pressure and wide reservoir applicability are major advantages for the miscible solvent, costs keep this process, one of the first tested miscible
techniques, an economic question mark. A schematic representation of the LPG process
is shown in figure 1.12.

Advantages
All contacted oil is totally displaced.
Low pressures are needed for miscibility.
The process is applicable in a wide range of reservoirs.
It can be used as a secondary or tertiary process.

Disadvantages
Process involves poor sweep efficiency, is best applied in steeply dipping beds.
Sizing of slug is difficult due to dispersion.
Slug materials are expensive.

ENRICHED GAS MISCIBLE PROCESS. In the enriched gas process, a slug of methane
enriched with ethane, propane, or butane (10 to 20% PV) and tailed by lean gas and/or
water is injected into the reservoir. As the injected gas contacts virgin reservoir oil, the
enriching components are stripped from the injected gas and absorbed into the oil. Continued injection of fresh enriched gas and the stripping of the light ends around the
wellbore form a zone rich in C_2 to C_4. One hopes that, if the injected gas is rich enough

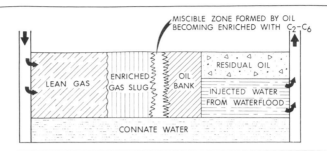

Figure 1.13 Enriched Gas Drive Process (Herbeck et al. 1976). Reprinted by permission of PETROLEUM ENGINEER International.

and in sufficient quantity, the enriched oil band around the wellbore at one point will become miscible with the injected gas.

Unlike liquid propane, which is miscible on first contact with the reservoir oil, this process relies on multiple contacts between oil and enriched gas to develop a miscible slug in situ. Although the cost of slug material is less than for LPG injection, higher pressures are required. Also, inherent in all gas drive processes is the lack of mobility and gravity control (areal and vertical sweep). Figure 1.13 is a schematic representation of the enriched gas process.

Advantages
The enriched gas process displaces essentially all residual oil contacted.
Miscibility can be regained if lost in the reservoir.
This process is lower cost than the propane slug process.
It develops miscibility at lower pressures than lean gas drive.
Large slug sizes minimize slug design problems.

Disadvantages
The process has poor sweep efficiency.
Gravity override occurs in thick formations.
Gas costs are high.
Viscous fingering leads to slug dissipation.

HIGH PRESSURE LEAN GAS MISCIBLE PROCESS. The lean gas process involves the continuous injection of high pressure methane, ethane, nitrogen, or flue gas into the reservoir. This process, like enriched gas, involves multiple contacts between reservoir oil and lean gas before a miscible bank is formed. However, unlike the enriched gas process where light components condense out of the injected gas and into the oil, intermediate hydrocarbon fractions (C_2 to C_6) are now stripped from the oil into the lean gas phase.

If the reservoir liquid is rich in intermediate fractions (C_2 to C_6), the leading edge of the gas front will become saturated with the reservoir oil's light ends and become mis-

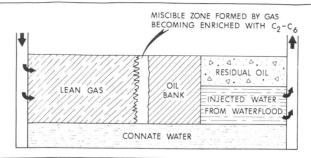

Figure 1.14 Lean Gas Drive Process (Herbeck et al. 1976). Reprinted by permission of PETROLEUM ENGINEER International.

cible with it. It is important to note that miscibility is attained not at the well face but at some distance away from the injection point. This distance, dependent on injection pressure and oil composition, may vary from a few feet to 100 feet before the lean injected gas has vaporized enough C_2 to C_6 to be miscible. This leaves a ring of residual saturation of stripped oil around the wellbore. Figure 1.14 shows the process as applied in the reservoir.

Advantages
The lean gas process provides a displacement efficiency approaching 100%.
Lean gas is less expensive than propane or enriched gas.
The process can regenerate miscibility if lost.
No slug sizing problems due to continuous injection occur.
Gas can be cycled and reinjected.

Disadvantages
The process has limited applicability because reservoir oil must be rich in C_2-C_6 components.
It involves high injection pressures.
Areal sweep efficiency and gravity segregation are poor.
Cost of natural gas is high; substitute gases require higher injection pressure.

1.4.2   Chemical Processes

MICELLAR POLYMER FLOODING. Micellar solutions are mixtures of surfactants (soap-like substances known as surface active agents), cosurfactants (alcohols for stability), electrolytes (salts such as sodium chloride or ammonium sulfate for viscosity and interfacial tension control), hydrocarbon (usually lease crude), and water. These solutions, designed on a field-by-field basis, are intended to displace reservoir oil and water miscibly.

Two different solution design concepts have developed. One injects a large pore vol-

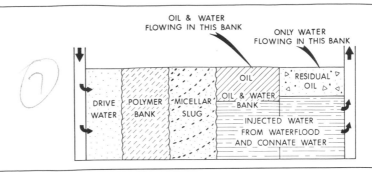

Figure 1.15 Micellar-Polymer Flooding (Herbeck et al. 1976). Reprinted by permission of PETROLEUM ENGINEER International.

ume of a low concentration surfactant solution, while the second concept uses a small pore volume (5 to 10% PV) of a high-surfactant-concentration micellar solution. Following the slug is a bank of polymer-thickened water for mobility control, and last, drive water is injected as in any waterflood.

Micellar fluids are not true miscible solutions. The fluids are designed to have low interfacial tension (1×10^{-5} dynes/cm) with the reservoir oil and water and, hence, do not displace 100% of the contacted oil. There are also many subclasses of micellar solutions, referred to as microemulsions, swollen micelles, fine emulsions, or soluble oils.

Figure 1.15 shows implementation of the process under tertiary conditions. Prior waterflooding has reduced in-place oil to a residual immobile saturation. Only water is produced until the oil-water bank reaches the producing well. This bank formation delay magnifies the economic risk involved of high front-end chemical costs with long-term income response (Gogarty 1978).

Advantages

This process involves high unit displacement and areal sweep efficiency.

Production technology is similar to waterflooding.

Gravity segregation is usually unimportant.

The process is applicable to wide range of reservoirs.

Disadvantages

Front-end chemical costs are high.

Performance prediction is poor due to mixing and dispersion of slug material.

Slug design process is sophisticated.

CAUSTIC FLOODING. Caustic or alkaline injection employs an in situ emulsification process. Caustic soda, sodium silicate, sodium carbonate, or sodium hydroxide is added to the injection water and mixes with residual oil in the reservoir. The crude oil must contain natural organic acids; most common are the naphthenic acids. As the alkaline-

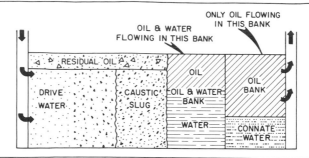

Figure 1.16 Caustic (alkaline) flooding.

injected water and acidic crude react, soaps are produced at the oil-water interface.

These soaps cause oil to be mobilized by one or more of the following mechanisms:

emulsification and entrainment,
emulsification and entrapment,
wettability reversal from oil-wet to water-wet,
wettability reversal from water-wet to oil-wet.

Figure 1.16 illustrates the process as a secondary recovery operation. Many times, polymer may be injected between the caustic slug and drive water to protect the integrity of the alkaline solution as well as to enhance sweep efficiency (Mayer et al. 1983).

Advantages
The process is relatively inexpensive to apply.
Mobility control is better than in gas injection processes.
The process is applicable to a wide range of crude oils.
Conversion from waterflooding to caustic flooding is easy.

Disadvantages
Corrosion potential may require coating of all piping, tanks, and tubing.
The process is not well suited for carbonate reservoirs.
Gypsum or anhydrite may precipitate in production wellbores.
Mixing and dispersion of alkaline solutions may cause poor response.

POLYMER FLOODING. Polymer floods are improved waterfloods that provide increased displacement efficiency by increasing displacing fluid viscosities. In addition, improvements in areal and vertical sweep can be achieved due to the improved mobility ratio by increasing the displacing fluid's viscosity and lowering its relative permeability through plugging.

The polymers are hydrated in water and form long chain molecules with molecular weights in the millions. The two most often used materials in polymer flooding are (1) polysaccharide biopolymers produced by the microorganism *Xanthomonas campestris*

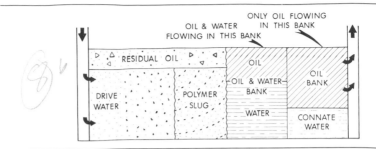

Figure 1.17 Polymer flooding.

and (2) polyacrylamides that are produced artificially by combining carbon, hydrogen, oxygen, and nitrogen into a basic unit called an *amine monomer*.

Polymer floods are conducted, as shown in figure 1.17, by injecting a slug of polymer solution chased with drive water. The process forms an oil bank and drives this bank as in a waterflood. Note that polymers provide no significant increase in mobilization efficiency. Their primary effect is to accelerate recovery and increase ultimate oil production (over a normal waterflood) due to improved areal and vertical sweep (Chang 1978).

Advantages
Areal and vertical sweep efficiency are increased.
Polymers are nontoxic and noncorrosive.
Polymer floods require similar production technology as waterfloods.
Use of polymers reduces producing water-oil ratios.

Disadvantages
Polymers are degradable either by chemical, bacterial, or shearing action.
Polyacrylamides require special surface handling.
Polysaccharides require filtration and bactericides.
Incremental oil recoveries may not warrant the extra front-end expense of polymers.

1.4.3   Thermal Processes

STEAM INJECTION PROCESS. The primary purpose of injecting heat into a reservoir is to reduce the oil viscosity and, thus, to improve the displacement efficiency. Crude expansion, steam distillation, and solvent extraction can also result in improved mobilization efficiency.

The process may involve *steam soak* (so-called huff-n-puff), where high quality steam is injected into a well at high injection rates. The well is then shut in, allowing the steam to soak the area around the wellbore. After a few days the well is put back on production until the producing oil rate declines to economic limits. The cycle is then repeated a number of times until no further response to steam is observed.

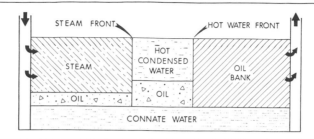

Figure 1.18 Steam Injection Process (Herbeck et al. 1976). Reprinted by permission of PETROLEUM ENGINEER International.

An alternate method using steam as a driving fluid is depicted in figure 1.18. *Steam drive* consists of continuous injection of steam, often of a lower quality than steam soak, into a pattern-type displacement similar to waterflooding. A steam zone forms around the injector and expands with continued injection. Ahead of the steam is a zone of condensed water formed by heat losses to the overburden, underburden, and the reservoir (Matthews 1983).

Advantages
Steam injection is a proved production technique where no other method may be feasible.
Steam generators can be fueled by produced oil or by gas or coal.
Front-end costs are moderate compared to chemical methods.
Displacement efficiency is high, recovering up to 60% of the original oil in place for steam drive.

Disadvantages
Ultimate recovery for steam soak is low, up to 10% of the original oil in place.
The process is limited by depth due to heat losses and high steam pressure.
Sand production is common.
Emulsion handling of produced fluids is necessary.
Good quality boiler-feed water is not always available.
Close spacing (typically 2.5 acre) is required.
Steam generator emissions cause air quality problems.

IN SITU COMBUSTION. In situ combustion also relies on crude oil viscosity reduction, expansion, distillation, and solvent extraction by the addition of heat to improve recovery. In contrast to steam injection, heat is generated within the reservoir rather than at the surface and is transported down the wellbore to the sand face.

The process is initiated by continuous injection of air (studies are examining pure oxygen injection; see Hansel et al. 1982; Hvizdos et al. 1982) into a well. Ignition of reservoir crude occurs either spontaneously or with the aid of a downhole heater. As in

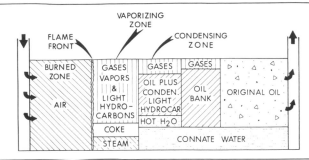

Figure 1.19 In Situ Combustion (Herbeck et al. 1976). Reprinted by permission of PETROLEUM ENGINEER International.

any combustion reaction, oxygen combines with the oil, forming carbon dioxide, carbon monoxide, and water, and releases heat. The combustion front moves forward through the reservoir only after burning all deposited fuel and extinguishes itself when the air flow at the front can no longer sustain combustion.

Figure 1.19 shows the various zones formed in an oil reservoir during the combustion process. Nearest the injection well is the *burned zone* through which the fire has progressed. All liquid has been removed from the rock, leaving only air-saturated pores. At the *flame front,* combustion of the deposited heavy fuel occurs at temperatures on the order of 600 to 1,200°F. Ahead of the flame, in the *vaporizing zone,* are combustion products, vaporized light hydrocarbons, and steam.

Next, as temperature lowers in front of the combustion zone, is the *condensing zone,* from which oil is displaced by miscible light hydrocarbons, hot waterflood, and combustion gas drive of reservoir crude. Displaced oil accumulates in the next zone, the *oil bank.* Here, immobile connate water, displaced oil, and some combustion gases occupy the pore space. Farther ahead lies the undisturbed reservoir that has not been affected by the fire (Chu 1982).

Advantages
In situ combustion is applicable to a wide variety of reservoirs up to 40° API.
The process involves more efficient heat generation than steam injection.
Displacement efficiency is high although some oil is burned.
Air is readily available.
The process may produce oil that is lighter than original oil.

Disadvantages
Design problems exist in controlling flame front.
Producing equipment can be damaged by heat.
Corrosion and emulsion handling are necessary.
Compression costs are high.
Gravity segregation may be a problem.
Noxious gas may be formed due to combustion.

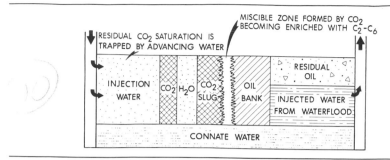

Figure 1.20 CO$_2$ Process (Herbeck et al. 1976). **Reprinted by permission of PETROLEUM ENGINEER International.**

1.4.4 **Carbon Dioxide Process.** Injection of CO$_2$ into oil reservoirs may initiate oil displacement by a number of mechanisms. Although not usually miscible with reservoir oil upon initial contact, CO$_2$ may create a miscible front similar to the lean gas process. Miscibility is initiated by extraction of significant amounts of heavier hydrocarbons (C$_5$ to C$_{30}$) by CO$_2$.

At different reservoir conditions, CO$_2$ displacement may resemble enriched gas drive; that is, CO$_2$ may saturate the reservoir fluids to an extent that the in situ swollen crude is miscible with trailing CO$_2$.

CO$_2$ may also be useful in heavy oil reservoirs where thermal methods may not be applied. CO$_2$-saturated crude oils exhibit moderate swelling, leaving fewer stock tank barrels of residual oil in place and reduced oil viscosities to a point where mobility ratios are drastically affected.

Figure 1.20 is a schematic representation of the miscible CO$_2$ process in horizontal reservoirs. A total of 12 to 40% of the original hydrocarbon pore volume may be injected and chased by foam or water. In steeply dipping beds, mobility control is not critical and the CO$_2$ slug can be chased by less expensive lighter gases (N$_2$, flue gas, etc.) (Holm 1982).

Advantages
Miscibility can be attained at low pressures.
Displacement efficiency is high in miscible cases.
This process aids recovery by solution gas drive.
It is useful over a wider range of crude oils than hydrocarbon injection methods.
Miscibility can be regenerated if lost.

Disadvantages
CO$_2$ is expensive to transport and not always available.
Poor sweep and gravity segregation can result under certain conditions.
Corrosion is increased.
Special handling and recycling of produced gas is necessary.

Table 1.4 Criteria for the Application of Selected EOR Methods

Screening Parameters	Steam Drive	In Situ Combustion	CO$_2$ Miscible	Surfactant/ Polymer	Immiscible CO$_2$
Viscosity, cp at reservoir conditions	NC	NC	<12	<20	100–1,000
Gravity, °API (California crudes)	>10 (>10)	10–45 (10–45)	>30 (>26)	>28 (>25)	10–25
Fraction of oil remaining in area to be flooded (before EOR), %PV	40[a]	50[a]	25	25	50
Oil concentration, B/AF porosity × oil saturation	>500 >.065	>400 >.05	NC	NC	>600 >.08
Depth, ft	<5,000[b]	>500	>3,000	NC (8,500)[b]	>2,300
Temperature, °F	NC	NC	NC	<200[b]	NC
Original bottomhole pressure, psi	NC	NC	>1,500	NC	>1,000
Net pay thickness, ft	>20	>10	NC	NC	NC
Permeability, md	NC	NC	NC	>20 (with polymer drive)	NC
Transmissibility (permeability × thickness/vicosity)	>100	>20	NC	NC	NC

Comments				
Porosity times thickness (high)	High dip preferred	Thin pay preferred	Homogeneous formation preferred	Thin pay preferred
10-acre spacing	Porosity times thickness high	High dip preferred	Low clay content	High dip preferred
Economic fresh water available	40-acre spacing	Homogeneous formation preferred	Porosity times thickness (high)	Homogeneous formation preferred
Economic fuel available	Low vertical permeability preferred	Porosity times thickness low	Prefer waterflood sweep 50%	Porosity times thickness low
High net to gross pay	High net to gross pay	Natural CO_2 availability		Natural CO_2 availability
Low clay content	Preferred temperature	Low vertical permeability in horizontal reservoirs		Low vertical permeability in horizontal reservoirs
No natural water drive	No natural water drive	No natural water drive	No natural water drive	No natural water drive
No major gas cap	No major gas cap	No major gas cap	No major gas cap	No major gas cap
Fractures not critical	No major fractures	No major fractures	No major fractures	No major fractures

[a] In portion of field to be flooded. Assuming 100% of area of reservoir contains 95% of remaining oil, the oil saturation for the total field becomes 42% of pore volume.

[b] Considered a constraint under current technology. NC = not a critical factor.

Source: *The Potential and Economics of Enhanced Oil Recovery*, FEA, Washington, D.C. (1976).

1.5   **APPLICATION SCREENING.** Each of the previously cited processes is economically applicable in a limited number of reservoirs. Table 1.4 summarizes the screening criteria that have been developed for various EOR techniques. The method of applying binary screens like these is to compare the properties of a given reservoir to the threshold values in the screen. If the reservoir fails any single criterion for a given process, then in theory, that process is considered inapplicable to the reservoir. While such an approach may be valid for purely technical screening, many of the criteria of table 1.4 are associated with the economic feasibility of the process and are therefore more subjective.

This, however, provides the reader with a view of the relative types of oils and reservoirs in which these processes may be applied. Keep in mind the screening values presented are inherently outdated with the addition of new technology and changing economics.

A better approach, presented in chapters 5 and 6, is to develop simple models that can be used to estimate oil recovery and EOR costs for a reservoir based on that reservoir's characteristics and then to determine the economics of that particular reservoir.

REFERENCES

"Annual Production Report," *Oil and Gas J.* (April 5, 1982) 139–159.

Chang, H. L.: "Polymer Flooding Technology—Yesterday, Today, and Tomorrow," *J. Pet. Tech.* (Aug. 1978) 1113–1128.

Chu, C.: "State-of-the-Art Review of Fireflood Field Projects," *J. Pet. Tech.* (Jan. 1982) 19–36.

DeGolyer, E. L., and MacNaughton, L. W.: *Twentieth Century Petroleum Statistics*, DeGolyer & MacNaughton, Dallas (1981).

Doscher, T. M., and Wise, F. A.: "Enhanced Oil Recovery Potential—An Estimate," *J. Pet. Tech.* (May 1976) 575–585.

Gogarty, W. B.: "Micellar/Polymer Flooding—An Overview," *J. Pet. Tech.* (Aug. 1978) 1089–1101.

Habermann, B.: "The Efficiency of Miscible Displacement as a Function of Mobility Ratio," *J. Pet. Tech.* (Nov. 1960) 264–268.

Hansel, J. G., Benning, M. A., and Fernbacher, J. M.: "Oxygen In-Situ Combustion for Oil Recovery: Combustion Tube Tests," paper SPE 11253 presented at the 1982 Eastern Regional Meeting, Washington, November 3–6, 1982.

Herbeck, E. F., Heintz, R. C., and Hastings, J. R.: "Fundamentals of Tertiary Oil Recovery," *Pet. Eng.* (Jan. 1976–Feb. 1977), nine-part series.

Holm, L. W.: "CO_2 Flooding—Its Time Has Come," *J. Pet. Tech.* (Dec. 1982) 2746–2756.

Hvizdos, L. J., Roberts, G. W., and Howard, J. V.: "Enhanced Oil Recovery via In-Situ Combustion: Test Results from the Forest Hills Field in Texas," paper SPE 11218 presented at the 57th Annual Fall Technical Conference, New Orleans, September 26–29, 1982.

Johnson, H. R.: *Outlook for Enhanced Oil Recovery*, DOE internal report, Bartlesville, Okla. (June 10, 1982).

Matthews, C. S.: "Steamflooding," *J. Pet. Tech.* (March 1983) 465–471.

Mayer, E. H., Berg, R. L., Carmichael, J. D., and Weinbrandt, R. M.: "Alkaline Injection for Enhanced Oil Recovery—A Status Report," *J. Pet. Tech.* (Jan. 1983) 209–221.

Moody, T.: "Where Oil & Gas Stand in the Energy & Ecology Dilemmas," *Oil and Gas J.* (Aug. 28, 1978) 185–190.

The Potential and Economics of Enhanced Oil Recovery, FEA, Washington, D.C. (1976).

"Winter's Legacy: Step-Up in Search for Fuel Supplies," *U.S. News & World Report* (Feb. 21, 1977) 19–20.

2   OIL RECOVERY MECHANISMS

This chapter provides a more detailed discussion of the oil recovery mechanisms introduced in the previous chapter. The overall recovery efficiency is a combination of individual process efficiencies, as shown in equation (2.1):

$$E_R = E_A \cdot E_V \cdot E_D \cdot E_M, \tag{2.1}$$

where

E_R = overall recovery efficiency,
E_A = areal sweep efficiency,
E_V = vertical sweep efficiency,
E_D = displacement efficiency,
E_M = mobilization efficiency.

Since several of these efficiencies are a function of the pore volumes injected (V_{pi}), care must be taken that each factor is evaluated at the same point in time so as not to overlap recovery provided by the other factors.

2.1   **DISPLACEMENT EFFICIENCY, E_D.** As defined previously, displacement efficiency is the fraction of moveable oil that has been displaced from the swept zone at any given time (for pore volumes injected). Thus,

$$E_D(V_{pi}) = \frac{S_{oi} - \overline{S}_o}{S_{oi} - S_{orp}}, \tag{2.2}$$

where $\overline{S}_o$ is the average oil saturation in the swept zone, S_{orp} is the ultimate residual oil to a given process, and S_{oi} is the initial oil saturation. For an *immiscible* displacement, $\overline{S}_o$ can be evaluated from the Buckley-Leverett frontal advance theory (1942).

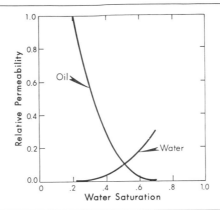

Figure 2.1 Typical relative permeability curves.

The fractional flow of water (or any immiscible displacing fluid) is given by

$$f_w = \frac{\dfrac{k_{rw}}{\mu_w}}{\dfrac{k_{rw}}{\mu_w} + \dfrac{k_{ro}}{\mu_o}} = \frac{1}{1 + \left(\dfrac{\mu_w\,k_{ro}}{\mu_o\,k_{rw}}\right)} \qquad (2.3)$$

[Gravitational and capillary forces have been ignored in eq. (2.3)]. Figure 2.2 shows fractional flow curves at various water-oil viscosity ratios for the relative permeability curves of figure 2.1. Examination of the fractional flow expression reveals that, taken alone, it is of little interest. This equation, coupled with the frontal advance formula, however, lays the foundation for theory defining the displacement process.

Buckley and Leverett (1942) showed that the velocity (v) of any given saturation through a linear, porous medium is proportional to the derivative of the fractional flow

$$(v)_{S_w} = \left(\frac{dx}{dt}\right)_{S_w} = \frac{q_t}{A\phi}\left(\frac{df_w}{dS_w}\right)_t, \qquad (2.4a)$$

or in finite difference terms

$$\Delta x = \frac{q_t \Delta t}{\phi A}\left(\frac{df_w}{dS_w}\right)_t, \qquad (2.4b)$$

where q_t is the total volumetric flow rate.

Examination of the frontal advance formula in this form shows that the linear advance of any saturation value can be determined as a function of time; that is, the advance can be calculated for a given time by taking the product of the group $q_t\Delta t/\phi A$ and the slope of the fractional flow–saturation curve at the saturation in question. Thus, the saturation distribution can be expressed as a continuous curve defined by a series of points determined by a series of slope values. Such a curve is shown in figure 2.3.

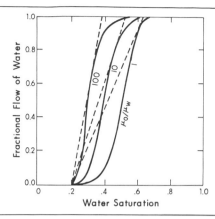

Figure 2.2 Fractional flow curves.

If this procedure is applied to a fractional flow curve such as that of figure 2.2, a point worthy of notice immediately arises. There are two regions of relatively high curvature in the fractional flow curve. If it is desired to obtain a saturation distribution in the reservoir at a given time, it is necessary to determine slopes and to solve the frontal advance equation. However, it is readily apparent that it is possible to have two slopes of equal value at two markedly different saturations in the displacing phase.

Thus, the solution of the frontal advance equation yields a distribution of the displacing phase such that two saturations exist at the same point. This physically impossible situation indicates that if the fractional flow curve possesses an inflection point, the displacement front will be a shock. Welge (1952) showed that the saturation at the shock front could be determined by constructing a line from the initial water saturation tangent to the fractional flow curve. The point of tangency would be the fractional flow of water and corresponding water saturation at the front at breakthrough:

$$\frac{S_{wf} - S_{wi}}{f_{wf}} = \frac{1}{\left(\dfrac{df_w}{dS_w}\right)_{f.}} \tag{2.5}$$

Here, S_{wf} is the water saturation at the shock, and f_{wf} is the corresponding fractional flow. The appropriate tangent lines are indicated in figure 2.2.

Welge (1952) also showed that the average water saturation ($\bar{S}_w$) behind the shock front at and after breakthrough is related to the pore volumes injected (V_{pi}) by

$$\bar{S}_w - S_{w2} = V_{pi} (1 - f_{w2}) \tag{2.6}$$

and

$$V_{pi} = \frac{1}{\left(\dfrac{df_w}{dS_w}\right)_{S_{w2}}}. \tag{2.7}$$

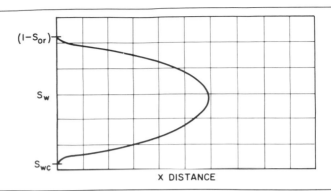

Figure 2.3 Water saturation as a function of distance for a given time.

where S_{w2} and f_{w2} are the producing water saturation and fractional flow respectively.

Equations (2.5), (2.6), and (2.7) provide the necessary relationships to evaluate $\bar{S}_o$ and, thus, E_D given in equation (2.2) as a function of cumulative injection. Figure 2.4 shows the displacement efficiencies calculated for the fractional flow curves of figure 2.2. Since E_D is the displacement in the swept zone only, it is a constant up to breakthrough. (Recovery increases prior to breakthrough due to an increase in the swept zone volume, not displacement efficiency.)

As indicated on figure 2.4, as the oil-water viscosity ratio decreases, displacement efficiency at a given total injection increases. While the displacement efficiency eventually reaches 1 in all cases, this occurs at essentially infinite total injection. If we take a 99% water cut ($f_w = 0.99$) as the economic limit for the process, the pore volumes injected at this point can be calculated using equation (2.7). One can see that the displacement efficiency at this limit decreases with increasing μ_o/μ_w while the total injection required increases.

The primary purpose of several EOR processes is to decrease the effective viscosity of the displaced fluid relative to the displacing fluid either by increasing the injected fluid's viscosity (polymer flooding) or by decreasing the oil viscosity (thermal processes, immiscible CO_2 injection) and thus both increase the economically attainable displacement efficiency and decrease the amount of injectant (and hence time) required to achieve this economic limit.

The previous discussion is for an *immiscible system*. Koval (1963) extended the fractional flow analysis to miscible systems by utilizing effective relative permeabilities to oil and solvent as a function of solvent saturation given by ($k_{ro} = 1 - S_s$) and ($k_{rs} = S_s$). These same relationships were used by Lantz (1970) and by Todd and Longstaff (1972) in their modifications of three-phase black oil simulators to model miscible displacements.

With k_{ro} and k_{rs} as defined, the fractional flow of solvent is given by

$$f_s = \frac{(\mu_o/\mu_s)S_s}{1 + S_s[(\mu_o/\mu_s) - 1]}. \tag{2.8}$$

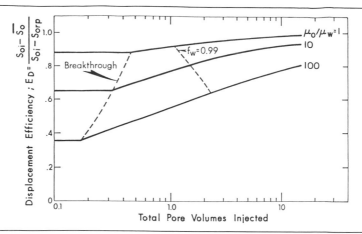

Figure 2.4 Displacement efficiencies calculated from immiscible fractional flow.

Equation (2.8) is obtained by substituting the expressions for the effective relative permeabilities into the fractional flow equation [eq. (2.3)] and replacing water properties with those of solvent. Note that the pure component viscosities are used in equation (2.8). This infers that miscible displacement behaves as if there were no mixing between the components. If there was complete and instantaneous mixing, the viscosity ratio would equal one and the fractional flow of solvent would be given by

$$f_s = S_s. \tag{2.9}$$

Koval (1963) found that neither of these extremes of oil solvent mixing matched the experimental data. However, if the viscosity ratio in equation (2.8) was replaced by an effective viscosity ratio, E, given by

$$E = \left[0.78 + 0.22 \left(\frac{\mu_o}{\mu_s} \right)^{1/4} \right]^4, \tag{2.10}$$

then the available experimental data could be matched with reasonable accuracy. Equation (2.10) has the form of the one-quarter power blending rule for mixture viscosities where the mixture is 22% solvent and 78% oil.

Figure 2.5 shows the calculated, miscible fractional flow curves for various viscosity ratios using Koval's E. Note that unlike the immiscible curves of figure 2.2, the miscible curves do not show an inflection point, so there is no shock front involved but a gradual grading of fluids in situ. Koval integrated equation (2.8) (using the effective viscosity ratio E) to determine total oil recovery as a function of pore volumes injected after breakthrough of solvent:

$$E_D = \frac{2(EV_{pi})^{1/2} - 1 - V_{pi}}{E - 1}. \tag{2.11}$$

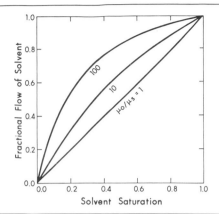

Figure 2.5 Fractional flow curves for miscible displacements calculated using Koval's E term.

The fractional flow of solvent *after breakthrough* is given by

$$f_s = \frac{E - \left[\dfrac{E}{(V_{pi})}\right]^{1/2}}{E - 1},\qquad(2.12a)$$

with the pore volumes injected at breakthrough defined at $f_s = 0$:

$$V_{pi} = \frac{1}{E}\qquad(2.12b)$$

and with the pore volumes injected for total recovery defined at $f_s = 1$:

$$V_{pi} = E.\qquad(2.12c)$$

Furthermore, the recovery (and, hence, displacement efficiency) prior to breakthrough is equal to the pore volumes injected prior to breakthrough.

Figure 2.6 shows the displacement efficiencies calculated for various oil-solvent viscosities using equation (2.11). Note that for $\mu_o/\mu_s = 1$, the displacement efficiency is identically 1, indicating that all of the oil would be displaced at solvent breakthrough. If fact, physical dispersion (not accounted for in Koval's treatment) would lead to solvent breakthrough somewhat before 100% oil recovery.

The fractional flow analysis is based on reservoir volumes. If we assume that the solvent has a formation volume factor of 0.5 bbl/MSCF and that the maximum economic producing GOR is 20 MSCF/bbl, the corresponding fractional flow would be about 0.91. Equation [2.12(a)] can then be used to estimate the pore volumes needed to reach this abandonment fractional flow. This point is indicated in figure 2.6.

Note that the displacement efficiency is again a strong function of viscosity ratio. Also note that the displacement efficiencies indicated in figures 2.4 and 2.6 are for the

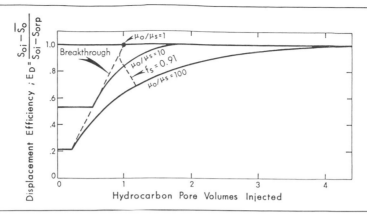

Figure 2.6 Displacement efficiencies for miscible displacements.

swept zone and do not include the effect of areal sweep (to be discussed later). Also, for the miscible case, it is assumed that a continuous injection of solvent is used. If a slug or water-alternated gas (WAG) injection scheme were used, the displacement efficiency relationships could be considerably different than those indicated.

2.2 **AREAL SWEEP EFFICIENCY, E_A.** As defined, areal sweep efficiency is the fraction of the total reservoir area that is invaded by the injected fluid and therefore includes not only the pattern sweep efficiency but also the fact that all of the reservoir area may not be well developed. Hence, areal sweep efficiency will be improved not only by improving the pattern sweep but also by infill or development drilling to enlarge the fraction of the reservoir that is being swept.

Pattern areal sweep efficiency is the fractional area of a pattern that has been invaded by injected fluid at a given injection volume and is a primary function of the fluid mobilities and pattern type. Craig (1971) has reviewed several studies of pattern sweep efficiency for various pattern types and has presented correlations of sweep versus mobility ratio.

The mobility of a fluid (λ) is defined as

$$\lambda_i = \frac{k_{ri}}{\mu_i}, \tag{2.13a}$$

and the mobility ratio, M, is the ratio of the displacing fluid's mobility to that of the displaced fluid such that

$$M = \frac{\lambda_{ing}}{\lambda_{ed}} = \frac{(k_r/\mu)_{ing}}{(k_r/\mu)_{ed}}. \tag{2.13b}$$

For *immiscible* fluids, the saturations at which the relative permeabilities are evaluated

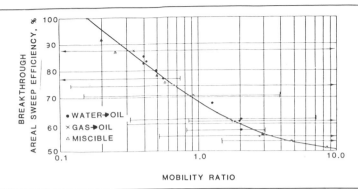

Figure 2.7 Areal sweep at breakthrough for a five-spot pattern (Craig 1971) ©SPE-AIME.

must be considered in determining M. Craig (1971) found that for the displaced fluid, the relative permeability should be evaluated at the initial saturation conditions, while for the displacing fluid, the relative permeability should be taken at the average saturation behind the flood front at breakthrough. This definition of mobility ratio resulted in a good correlation of breakthrough sweep efficiency for water-oil, gas-oil, and miscible fluid systems as indicated in figure 2.7. In addition, for *miscible* displacements, the relative permeability of displaced and displacing fluids is assumed to equal unity, reducing the mobility ratio equation to the viscosity ratio.

Applying Craig's definition of M to the fractional flow curves of figure 2.2 and to the areal sweep efficiencies correlation of figure 2.8, one can calculate the breakthrough sweeps given in table 2.1 for various viscosity ratios. These results indicate the ameliorating effect of relative permeability on mobility ratio and hence on areal sweep. A one hundred–fold increase in oil-water viscosity ratio resulted in a less than tenfold increase in mobility ratio and less than a 35-percentile decrease in breakthrough areal sweep.

The pore volumes of injected water needed for breakthrough can be estimated by

$$(V_{pi_{bt}}) = E_{A_{bt}} (\overline{S}_{w_{bt}} - S_{wi}), \tag{2.14a}$$

while additional area sweep values after breakthrough (for enclosed five spots) may be estimated using

$$E_A(V_{pi}) = E_{A_{bt}} + 0.63117 \cdot \log_{10}(V_{pi}/V_{pi_{bt}}) \tag{2.14b}$$

where E_A cannot be greater than 1 (Stahl 1960).

Note that like displacement efficiency, areal sweep efficiency also increases with continued injection after breakthrough. Figure 2.8 shows the areal sweep efficiency as a function of M for a five-spot pattern at various producing cuts. The areal sweeps at a 98% water cut for the various viscosity ratios are also given in table 2.1. The areal sweeps for all cases are essentially 100% at this point, indicating that even at viscosity ratios as high as 100, very good sweep efficiencies can be obtained in immiscible displacements. Note that this does not mean that nearly 100% oil recovery will be achieved. In order to determine oil recovery, the displacement, vertical sweep, and

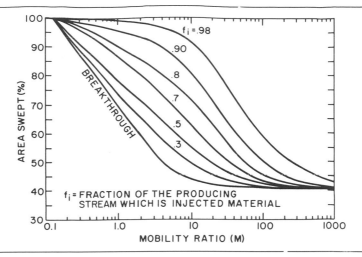

Figure 2.8 Sweep efficiency as a function of mobility ratio for a five-spot pattern at various producing cuts (Caudle and Witte 1959) ©SPE-AIME.

mobilization efficiencies must also be taken into account. From figure 2.4, $E_D = 0.64$ at $f_w = 0.99$ for $\mu_o/\mu_w = 100$. From table 2.1, $E_A \cong 0.97$ at $f_w = 0.98$ for $\mu_o/\mu_w = 100$.

For $S_{oi} = 0.8$ and $S_{orp} = S_{or}$ after a waterflood $= 0.3$ (fig. 2.1),

$$E_M = \frac{0.8 - 0.3}{0.8} = 0.625 \text{ (with no swelling).}$$

Assuming 100% vertical sweep and an areal sweep of 97%, the estimated overall recovery is *only:*

$$
\begin{aligned}
E_R &= E_A \cdot E_V \cdot E_D \cdot E_M \\
&= (0.97)(1)(0.64)(0.625) \\
&= 0.388, \text{ or } 38.8\% \text{ OOIP (original oil in place).}
\end{aligned}
$$

As pointed out earlier, for a secondary, miscible displacement, the mobility ratio is

Table 2.1 Mobility Ratios and Areal Sweep Efficiencies for Immiscible Displacements

$\dfrac{\mu_o}{\mu_w}$	$\overline{S}_w$	$k_{rw} (\overline{S}_w)^*$	M	$(E_A)_{bt}$	$(E_A)_{fw} = 0.98$
1	0.670	0.260	0.26	0.91	1.00
10	0.525	0.110	1.10	0.66	0.99
100	0.380	0.025	2.50	0.57	0.97

*$k_{ro} = 1$

equal to the oil-solvent viscosity ratio. For a viscosity (mobility) ratio of 100, we get from figure 2.8 that E_A at breakthrough is 0.40, and at a 0.98 fractional flow, M is only 0.58. These areal sweep efficiencies are much lower than those shown in table 2.1 for an immiscible displacement.

In order to improve the areal sweep efficiency of miscible displacements, Caudle and Dyes (1958) investigated the simultaneous injection of solvent and water. The objective was to reduce the solvent mobility in the swept zone by decreasing its relative permeability. They recommended injecting a slug of solvent large enough to maintain a solvent bank between the oil and advancing waterfront and driving this bank through the reservoir by simultaneous injection of water and solvent (or some fluid miscible with the solvent). The water-solvent ratio is designed so that the relative velocities of the two phases are equal. This condition is met when

$$v_s = \frac{k_{rs}}{\mu_s S_s} = \frac{k_{rw}}{\mu_w (S_w - S_{wc})} = v_w. \tag{2.15}$$

For $\mu_w = 0.3$ cp, $\mu_s = 0.03$ cp, and using the relative permeability curves of figure 2.1, the relative velocities calculated from equation (2.15) are shown in figure 2.9. Equality of relative velocities is achieved at a water saturation of 0.6. At a water saturation of 0.6 and a critical water saturation of 0.2, the water-solvent injection ratio (in reservoir volumes) should equal $(S_w - S_{wc})/S_s$ or unity in this case. Alternatively, the use of foam to reduce mobility and thereby to increase areal sweep is currently being studied for both CO_2 and steam injection systems.

For a *tertiary, miscible* process, water is being displaced by oil (if a free oil bank is developed) or, more likely, by some mixture of oil and solvent. Thus, the areal sweep is determined by the mobility of the water ahead of the front and the effective mobility of the solvent-oil mixture behind the displacement front. While the evaluation of the water mobility is straightforward, the fact that the viscosity of the nonaqueous phase behind the front can vary from that of oil to that of pure solvent makes it difficult to determine the effective mobility ratio. However, let us consider two extremes: (1) that the water displacement and sweep efficiency can be characterized by oil-displacing water and (2) that the water displacement is characterized by CO_2-displacing water.

Figure 2.10 shows the fractional flow curves for these two cases, assuming $\mu_w = \mu_o = 0.3$ cp, $\mu_s = 0.03$ cp. Also shown are the Welge tangent lines for the process. Note that for these cases, the initial water saturation and fractional flow conditions are $S_w = 1 - S_{orw}$ for a waterflood and $f_w = 1$. Therefore, the tangent lines start at this condition.

For oil displacing water, the average water saturation behind the front is 0.375, while $\bar{S}_w = 0.520$ for solvent displacing water. Table 2.2 shows the estimated mobility ratios and areal sweeps for these cases. Also included in figure 2.10 and table 2.2 are the results for a mixture of 22% solvent and 78% oil displacing water. This corresponds to the effective viscosity developed by Koval for miscible displacements.

While none of the cases can be said to represent the actual process, the relatively low mobility ratios (<3) in table 2.2 indicate that the areal sweep efficiency of a tertiary miscible displacement is higher than might be expected when only the viscosities of the various fluids are considered. The fact that the areal sweep at breakthrough is relatively high does not mean that rapid oil and solvent breakthrough does not occur. The pore volumes injected at breakthrough depend not only on the areal sweep but also the swept

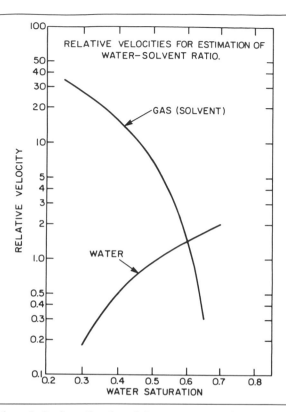

Figure 2.9 Relative velocity for estimation of the proper water-solvent ratio.

zone displacement efficiency. Breakthrough volumes are given by, assuming 100% vertical sweep,

$$(V_{pi})_{bt} = E_{Abt}(S_{wi} - \bar{S}_{wbt}) \tag{2.16}$$

Thus, for solvent displacing water, breakthrough would occur at about 0.10 PV injected while, if oil is the effective displacing fluid, breakthrough occurs at about 0.22 PV.

Table 2.2 Estimated Mobility Ratios and Areal Sweeps for Tertiary Miscible Displacements

Displacing Fluid	μ	$\bar{S}_w$	$k_{ro}(\bar{S}_w)$	M	$(E_A)_{bt}$
Oil	0.3	.375	0.34	1.13	0.68
Solvent	0.03	.520	0.08	2.67	0.55
Mixture	0.16	.415	0.24	1.50	0.64

Note: $\lambda_{water} = \dfrac{0.3}{0.3} = 1$.

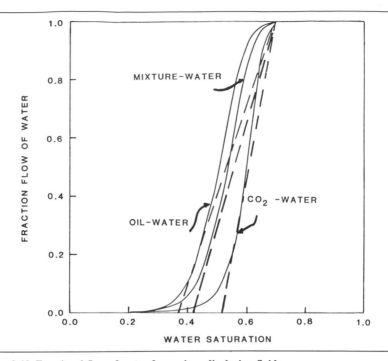

Figure 2.10 Fractional flow of water for various displacing fluids.

2.3   **VERTICAL SWEEP EFFICIENCY, E_v.** Vertical sweep efficiency is the fraction of the vertical section of a reservoir that has been contacted by the injected fluid. Maximum sweep is obtained when the displacement front proceeds through the reservoir as a plane perpendicular to the bedding plane. Any forces that act to distort this plane will reduce vertical sweep, and conversely, any forces that serve to bring a distorted front toward perpendicular will increase vertical sweep.

2.3.1   **Gravity Override**

DIPPING RESERVOIRS. The two factors that are considered most important in determining vertical sweep are gravity override and vertical heterogeneity. Hawthorne (1960) considered the case of gravity override (or tongueing) for situations where the displacement efficiency was essentially one and where stabilized interfaces could develop. Consider the immiscible displacement shown in figure 2.11 with gas displacing oil downdip. Here, α is the dip angle and β is the angle of the interface relative to horizontal. The interface is stable (i.e., β will be a constant throughout the displacement) if

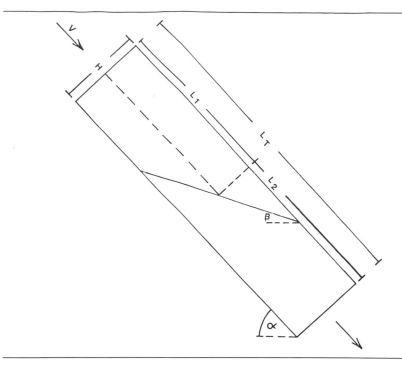

Figure 2.11 Geometry of Hawthorne's system (1960).

$$v < v_c = \frac{(\rho_o - \rho_g)}{\frac{\mu_o}{k_o} - \frac{\mu_g}{k_g}} g \sin \alpha. \tag{2.17}$$

Here, v is the fluid velocity and v_c is the critical rate. Rho is the fluid density while μ is the fluid viscosity. Note that this equation is equally applicable for updip displacements. However, for this case, the velocity v is negative. Also, since a unit displacement efficiency is assumed, the relative permeabilities are evaluated at their endpoints; thus, k_o is evaluated at initial oil saturation while k_g is calculated at the residual oil saturation.

Dumoré (1964) developed a similar relationship for the critical rate in a miscible displacement by assuming $k_o = k_g = k$ so that

$$(v_c)_{misc} = \frac{\rho_o - \rho_g}{\mu_o - \mu_g} (k \, g \sin \alpha). \tag{2.18}$$

In addition, Dumoré showed that, if the transition zone between the solvent (gas) and the oil is considered, an additional stability criteria (called the *stable rate*) can be developed. At rates less than the stable rate, no viscous fingers develop. The displacement is totally stable and is dominated by a single, gravity tongue. At rates between the stable and critical rate, part of the transition zone forms viscous fingers and the displacement is only partially stable. For rates exceeding the critical rate, the process is totally unstable and vertical sweep is dominated by viscous fingering rather than gravity override.

The stable rate is dependent on the relationship between mixture composition, density, and viscosity. If linear blending of densities and logarithmic blending of viscosities are assumed,

$$\rho_m = C_g\rho_g + (1 - C_g)\rho_o \qquad (\rho_g < \rho_o) \qquad (2.19a)$$

and

$$ln(\mu_m) = C_g ln(\mu_g) + (1 - C_g)ln(\mu_o) \qquad (\mu_g < \mu_o), \qquad (2.19b)$$

where C_g = concentration of gas (solvent) in the mixture, then the stable rate is given by

$$v_{st} = \frac{\rho_o - \rho_g}{\mu_o(ln\mu_o - ln\mu_g)} k\, g\, \sin\alpha \qquad (2.20a)$$

or

$$\frac{v_{st}}{v_c} = \frac{1 - \dfrac{\mu_g}{\mu_o}}{ln(\mu_o/\mu_g)}. \qquad (2.20b)$$

Figure 2.12 shows results of experiments conducted by Dumoré for vertical, miscible displacements. As expected, breakthrough recovery decreases as the rate increases above the stable rate and viscous fingers form. But the severest reduction in recovery occurs at rates above the critical rate where viscous fingers dominate.

Gardner and Ypma (1982) have postulated that not only density, viscosity, permeability, and dip angle determine the stabilized flood rate above which viscous fingers appear, but that dispersion of solvent into the oil phase must be accounted for. For unstable displacements, as solvent tries to finger through the displaced oil, it disperses into the oil normal to finger growth and reduces the size of the fingers. That is to say, stabilized rates as predicted by Dumoré may be low. High solvent dispersion should help retard the growth of viscous fingers and allow floods to be run at high rates without viscous fingers appearing.

If equation (2.17) is not met, the interface is not stable and a finger of injected fluid will appear and grow without limit until the interface is parallel to formation dip. However, if equation (2.17) is satisfied, the angle can be determined by

$$\tan\beta = \frac{v\cos\alpha\left(\dfrac{\mu_o}{k_o} - \dfrac{\mu_g}{k_g}\right)}{g(\rho_o - \rho_g) - v\sin\alpha\left(\dfrac{\mu_o}{k_o} - \dfrac{\mu_g}{k_g}\right)}. \qquad (2.21)$$

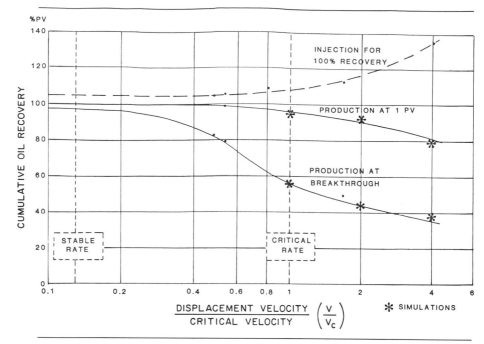

Figure 2.12 Experimental results of linear solvent drives in a vertical downward direction (Dumoré 1964) ©SPE-AIME.

Equation (2.18) can be used to estimate breakthrough time for a stabilized front. At Q_i barrels of gas injection, L_1 is the length of the gas zone, L_2 is the oil zone, while h is the thickness.

$$L_1 h(S_{oi} - S_{or}) = Q_i. \tag{2.22}$$

Also,

$$L_2 = \frac{1}{2} \frac{h}{\tan (\alpha - \beta)} \tag{2.23}$$

At breakthrough,

$$L_1 + L_2 = L_t$$

so that

$$\frac{Q_i}{h(S_{oi} - S_{or})} + \frac{1}{2} \frac{h}{\tan (\alpha - \beta)} = L_t. \tag{2.24}$$

The pore volume injected at this point is

$$V_{pi} = \frac{Q_i}{L_t h}$$

so that

$$(V_{pi})_{bt} = (S_{oi} - S_{org}) \left[1 - \frac{1}{2} \frac{\frac{h}{L_t}}{\tan (\alpha - \beta)} \right]. \tag{2.25}$$

Since the areal sweep and displacement efficiency are assumed both to be one for this case, the vertical sweep at breakthrough is equal to the recovery at breakthrough. Therefore,

$$(E_V)_{bt} = 1 - \frac{1}{2} \frac{\frac{h}{L_t}}{\tan (\alpha - \beta)} \tag{2.26}$$

or

$$(E_V)_{bt} = 1 - \frac{1}{2} \frac{\frac{h}{L_t}}{\tan \alpha} \left(\frac{1}{1 - \frac{v}{v_c}} \right) \qquad \left(\frac{v}{v_c} < 1 \right). \tag{2.27}$$

Equation (2.27) provides a method for estimating vertical sweep at breakthrough for a stable interface ($v/v_c < 1$). This equation indicates that as the thickness/length of a reservoir decreases and/or the displacement rate decreases, the vertical sweep increases. From the definition of v_c [eq. (2.17)], it can be seen that for $\alpha = 0$, $v_c \cong 0$ and, therefore, that no stable interface can develop. This approach, then, cannot be used to estimate breakthrough vertical sweep efficiency in horizontal reservoirs.

HORIZONTAL RESERVOIRS. Craig et al. (1957) studied vertical sweep efficiency for horizontal floods at low rates where the displacement could be characterized as stable (no viscous fingers with vertical sweep dominated by a single gravity tongue). Figure 2.13 shows their results for a linear, uniform system using miscible fluids. For such a system, areal and displacement efficiencies are very nearly one, so that the volumetric sweep efficiency is essentially the vertical sweep efficiency. The abscissa on figure 2.13 represents the ratio of viscous to gravity forces, while M is the mobility ratio of the displacement. Craig's work shows that high flood velocities increase vertical sweep since less time is available for the lighter solvent to migrate to the top of the formation. These results also indicate that for mobility ratios greater than one, gravity override can result in a drastic decrease in oil recovery.

Consider a *secondary*, miscible displacement with the following data.

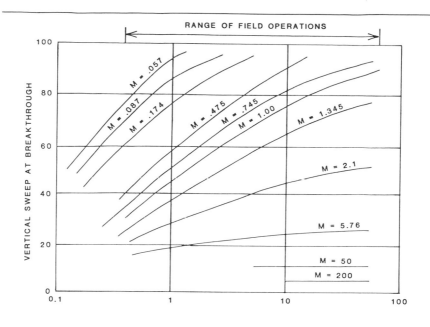

$$\left(\frac{\Delta P_h}{\Delta P_v}\right) = \frac{q_i \mu_o L}{k_x g \Delta \rho h}$$

Figure 2.13 Vertical sweep efficiency at breakthrough for linear uniform systems (horizontal) (Craig et al. 1957) ©SPE-AIME.

Velocity	1 ft/D
L	1,500 ft
ρ_o	40 lb/ft^3
ρ_{CO2}	28 lb/ft^3
μ_o	0.3 cp
μ_{CO2}	0.05 cp
k_x^*	1,000 md

Note: If the vertical permeability, k_z, is not equal to the horizontal permeability, k_x, then $\sqrt{k_z \cdot k_x}$ may be substituted for k_x.

$$\frac{(\Delta P_h)}{(\Delta P_v)} = \frac{q_i \mu_o L}{k_x g \Delta \rho h}$$

$$= \frac{1 \text{ ft/d } 0.3 \text{ cp } 1{,}500 \text{ ft}}{(1{,}000 \text{ md})(32.174 \text{ ft/sec}^2)(12 \text{ lbm/ft}^3)(30 \text{ ft})}$$

$\cdot (1 \text{ d}/24 \text{ hr})(2.42 \text{ lbm/ft-hr-cp})$

$\cdot (1 \text{ md}/1.06232 \cdot 10^{-14} \text{ft}^2)$

$\cdot (1 \text{ hr}/3{,}600 \text{ sec})^2$

$$\frac{(\Delta P_h)}{(\Delta P_v)} = 28.4$$

$$M = \frac{\mu_o}{\mu_{CO_2}} = 6.$$

Note that vertical sweep efficiency in horizontal systems improves as velocity increases. This increase in sweep occurs because less time is available for the injected fluid to segregate to the top of the formation and override significant portions of the reservoir. In dipping beds the opposite is true. Slower floods allow the lighter fluids to segregate with a subsequent increase in vertical sweep. For this case, the viscous to gravity force ratio $(\Delta P_h / \Delta P_v)$ is 28.44 and the mobility ratio is 6, resulting in a break-through vertical sweep efficiency of about 25%.

In *tertiary* recovery, Stalkup (1983) suggests that a rough estimate of vertical sweep-out can be made from figure 2.13 by substituting

$$\frac{1}{\sqrt{k_x k_z}\left(\dfrac{k_{roi}}{\mu_o} + \dfrac{k_{rwi}}{\mu_w}\right)}$$

for μ_o/k_x, where k_{roi} and k_{rwi} are the oil and water relative permeabilities at the start of solvent injection and $\Delta\rho$ is the density difference between solvent and water. A rough approximation of mobility ratio can be made using the techniques described in Section 2.2.

Blackwell et al. (1960) suggested using simultaneous water-solvent inject to improve vertical sweep in miscible displacements by reducing the effective gas mobility. Assuming complete segregation of water and solvent, they developed the following relationship for the injected water-solvent ratio (in reservoir volumes) so that the water and solvent fronts advance at the same rate:

$$\frac{Q_w}{Q_g} = \frac{(1 - S_{rg} - S_{wc}) - (1 - S_{wc})\dfrac{k_w(S_{rg})\mu_g}{k_g(S_{wc})\mu_w}}{S_{org}}, \tag{2.28}$$

where S_{rg} is the residual solvent saturation, $k_w(S_{rg})$ is the water permeability at residual solvent saturation, and $k_g(S_{wc})$ is the solvent permeability at connate water.

The effective total mobility in the water-solvent zone is given by

$$M_{eff} = \frac{1 + \dfrac{Q_w}{Q_g}}{\dfrac{k_w(S_{rg})\mu_g}{k_g(S_{wc})\mu_w} + \dfrac{Q_w}{Q_g}} \frac{k_w(S_{orw})\mu_o}{k_o(S_{wc})\mu_w}. \tag{2.29}$$

Laboratory results showed that simultaneous injection of water and solvent at a ratio of 3.3 to 1 resulted in a recovery of 83% at 1 HCPV injected as opposed to about 70% for straight solvent injection. A waterflood gave a recovery of about 68% for the same system.

In *tertiary* miscible displacements, besides solvent-oil override, underride of the non-aqueous phase by water is important. Warner (1977) conducted simulation studies of tertiary CO_2 floods that indicated definite gravity segregation of water and CO_2. His results also indicated that oil recovery could be increased by the use of simultaneous CO_2 and water injection, as well as the WAG process.

Later work by Fayers et al. (1981) reviewed the aspects of CO_2 gravity segregation. They studied four principal cases: (1) continuous CO_2 injection; (2) CO_2 injection as a slug for 500 days, followed by chase water; (3) CO_2 injection as a slug for 1,000 days, followed by chase water; and (4) simultaneous injection of CO_2 into the lower half and water into the upper half for 1,000 days, followed by chase water. Case 1 achieved a recovery of 45% of the target oil after 3,000 days of CO_2 injection. Case 2 achieved 32% recovery after 1,000 days of chase water. Case 3 gave 42% recovery after 1,200 days of chase water. Simultaneous injection in case 4 yielded 40% recovery after 700 days of chase water and appeared to be the most economical in terms of CO_2 utilization.

Youngren and Charlson's (1980) history match analysis of the Little Creek CO_2 pilot indicated that the oil recovery was dominated by gravity segregation and crossflow of oil. It should be pointed out that both the Fayers et al. (1981) and this study dealt with sandstone-type reservoirs characterized by relatively high permeabilities and very good vertical transmissibility. A study by Bilhartz et al. (1978) of the Willard Unit CO_2 pilot concluded that, for this carbonate reservoir, gravity segregation was relatively unimportant and that stratification of the reservoir was the dominant effect. Kane (1979) reports similar conclusions for the SACROC field (also a carbonate) in that the oil recovery was dominated by vertical and areal heterogeneity.

2.3.2   **Vertical Heterogeneity.** The effect of stratification or vertical heterogeneity on vertical sweep and, hence, oil recovery is discussed extensively by Craig (1971). The most commonly used method of characterizing vertical heterogeneity is the permeability variation coefficient proposed by Dykstra and Parsons (1951), which makes use of the fact that rock permeabilities are usually log normally distributed. In determining the Dykstra-Parsons coefficient (V_{DP}), the core-derived permeability values are arranged in descending order, and the percentage total number of permeability values exceeding each entry is computed and plotted on log-normal probability paper. The best straight line through the points is used to determine $\bar{k}$ (permeability at 50%) and k_σ *(permeability at 84.1%)*. V_{DP} is then given by

$$\frac{\bar{k} - k_\sigma}{\bar{k}}. \tag{2.30}$$

This procedure is illustrated in figure 2.14. The Dykstra-Parsons V_{DP} does not take into account the ordering or existence of discrete layers within the reservoir; that is, a randomly heterogeneous reservoir can give the same V_{DP} as one with definite stratification.

Various methods have been developed to predict oil recovery from vertically heterogeneous reservoirs. These methods assume that the reservoir is made up of a stack of noncommunicating layers, and thus, the oil recovery and water cut is based on the sum of the individual layer behavior. The Stiles (1949) model assumes that injection into

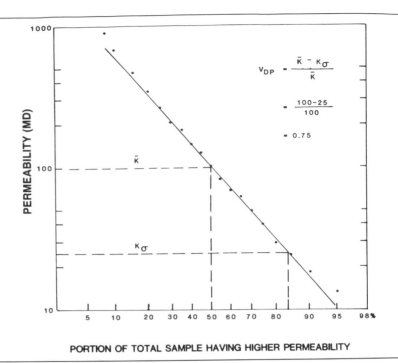

$$V_{DP} = \frac{\bar{K} - K_\sigma}{\bar{K}}$$

$$= \frac{100 - 25}{100}$$

$$= 0.75$$

PORTION OF TOTAL SAMPLE HAVING HIGHER PERMEABILITY

Figure 2.14 Definition of Dykstra-Parsons Permeability Variation Coefficient (1950).

each layer is proportional only to the layer's permeability-thickness product and that pistonlike displacement occurs.

Dykstra and Parsons also developed a correlation of waterflood recovery as a function of mobility ratio and permeability variation. The results of linear corefloods were used to calculate recoveries for a layered linear model with no crossflow. Figure 2.15 shows the vertical sweep efficiency (as calculated by the Dykstra-Parsons method) at a producing water-oil ratio (WOR) of 100 for various mobility ratios and V_{DP}.

The discussion of vertical sweep has so far focused on two extremes: the homogeneous, gravity-dominated reservoir and the heterogeneous, zero crossflow system. In fact, very few, if any, reservoirs are truly represented by these two extremes. Heterogeneity and crossflow exist to some extent in all reservoirs. However, the theoretical analysis of such systems is nearly impossible. Currently, numerical simulation models are the principle tool available to analyze these mixed systems.

2.4   **MOBILIZATION EFFICIENCY, E_M.** Mobilization efficiency is the fraction of the oil in place at the start of a process that can be recovered at 100% areal, vertical, and displacement efficiency and, as such, represents the maximum target for any EOR process. We define E_M as

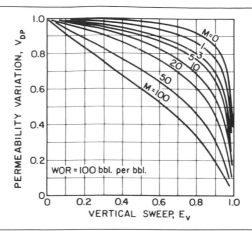

Figure 2.15 Effect of permeability variation on vertical sweep efficiency at a WOR of 100 (Dykstra and Parsons 1950).

$$\frac{S_{oi}/B_{oi} - S_{orp}/B_{of}}{S_{oi}/B_{oi}}, \tag{2.31}$$

where

S_{oi} = oil saturation at start of project,
B_{oi} = oil formation volume factor at start of project,
S_{orp} = residual oil to process,
B_{of} = oil formation volume factor at end of process.

The mobilization efficiency is determined primarily by the ratio of capillary to viscous forces (on a microscopic level) and by interphase mass transfer.

For an immiscible displacement, the residual oil saturation is dependent on the ratio of capillary to viscous forces, characterized by the capillary number:

$$N_c = \frac{v\mu}{\sigma}, \tag{2.32}$$

where

v = displacing fluid velocity,
μ = displacing fluid viscosity,
σ = interfacial tension.

In addition to capillary number, the residual oil to a given process will depend on pore structure and wettability characteristics of the porous system. Figure 2.16 shows the effect of capillary number on residual oil saturation obtained by various investigators. From this figure we can see that to reduce the residual oil saturation significantly below

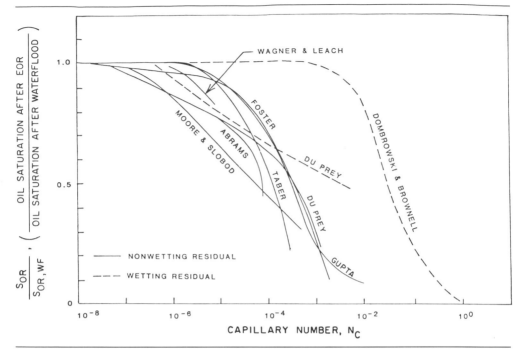

Figure 2.16 Recovery of residual oil versus capillary number (Stegemeier 1977; Gupta and Trushenski 1979).

that of a waterflood requires an increase in capillary number of four orders of magnitude or more. The velocity in a reservoir is given by Darcy's law to be

$$v = \frac{k}{\mu} \frac{\Delta P}{\Delta L}. \tag{2.33}$$

With this relationship, the capillary number becomes

$$N_c = \frac{k\dfrac{\Delta P}{\Delta L}}{\sigma}. \tag{2.34}$$

In a reservoir, it is not practical to increase significantly the pressure differential between production and injection wells (ΔP) over that of a waterflood. Infill drilling can feasibly reduce the interwell distance (ΔL) by no more than a factor of two to four. Therefore, the only practical method of increasing the capillary number the four orders of magnitude required to reduce significantly S_{orp} and hence to increase E_M over a waterflood is to reduce the interfacial tension. Micellar polymer floods achieve this reduction in σ by the use of surface active agents that can achieve interfacial tensions between the aqueous and nonaqueous phases as low as 10^{-4} dynes/cm. Caustic flooding achieves low interfacial tensions through the in situ generation of surface active agents.

In secondary miscible floods, the displacing agent is miscible with the oil, and therefore, the interfacial tension is zero, which means there is no tendency for the miscible fluid to trap oil and (S_{orp}) approaches 0. In fact, most miscible floods are actually multiple-contact miscible displacements that will leave some small residual oil saturation (3 to 10%).

In tertiary miscible floods, the oil is initially shielded from the injected solvent by the water. While there is zero interfacial tension between the injected fluid and oil, the aqueous-nonaqueous interfacial tension is relatively unaffected. Therefore, in order for a miscible fluid to displace waterflood residual oil, it must first displace the water in order to contact the oil.

Several investigators have shown that in the presence of mobile water $(S_w > S_{wc})$, oil is trapped as a fraction of oil that is flowable. The rest resides in locations that are blocked by water. The fraction of oil that is trapped depends on the water saturation. Shelton and Schneider (1975) showed that it could be determined from the envelope between the drainage and imbibition relative permeability curves (see fig. 2.17). Displacements at constant water saturations by Raimondi and Torcaso (1964) indicated that this trapped oil could not be displaced without reducing the water saturation.

In order to understand the potential magnitude of this effect, consider the fractional flow curves of figure 2.10 and the corresponding average swept zone saturations given in table 2.2. For oil displacing water at the front, the average swept zone water saturation is 0.375. From figure 2.17, this gives a trapped oil saturation of 7% of S_{orw}, or 0.021. The corresponding mobilization efficiency at breakthrough (assuming $B_{oi} = B_{of}$) would be 0.93. However, for solvent displacing water, $S_{ot} = 0.195$ and $E_M = 0.350$, while for the Koval mixture, $S_{ot} = 0.040$ and $E_M = 0.865$. Thus, the trapping phenomenon can have a significant effect on the timing and magnitude of ultimate oil recovery. Results of solvent flood simulations with and without the trapped effect are shown in figure 2.18.

The trapped oil effect has appeared to be quite severe in solvent flooding strongly water-wet laboratory sandstones. However, trapping of oil does not seem to be as severe in rocks of intermediate or oil wettability. Also, studies by Stalkup (1970) have indicated that much of the trapped oil, whether trapping is by water blockage or low permeability bypassing, may be recoverable by diffusion into the flowing stream. The fact that CO_2 is soluble in water should increase the oil recovery resulting from this diffusion. Furthermore, Stalkup's (1970) results indicate that the effect of oil trapping by water will be more significant in laboratory core experiments than in field scale operations.

In fact, oil recovery by CO_2 injection as a multiple-contact miscible material does not appear to be as affected by mobile water remaining after a waterflood. Early research of oil-trapping effects involved long-core tests using a non-water-soluble solvent that was first-contact miscible with the in-place oil. Tiffin and Yellig's (1982) work with CO_2 core displacements revealed that mobile water from a previous waterflood changed neither the mass transfer process to develop miscibility nor the overall recovery with CO_2 injection. They also noted, however, that water injected simultaneously with CO_2 in water-wet, postwaterflooded cores did trap significant amounts of oil and interfered with the development of miscibility. However, Stalkup (1980) has not found significant oil trapping in preserved carbonate cores when water and CO_2 are injected in a one-to-one volumetric ratio. It appears that in-place water does not affect recovery by the CO_2

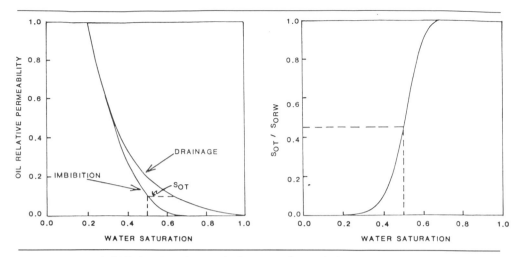

Figure 2.17 Estimation of trapped oil amounts from relative permeabilities.

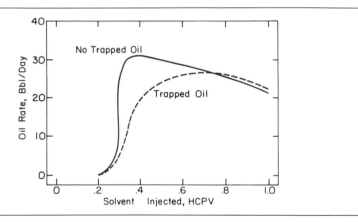

Figure 2.18 Comparison of oil recovery with and without trapped oil effects.

process but that there may be some optimum value of injected water/CO_2 ratio to minimize the oil-trapping effects.

These discussions have dealt with mechanisms that affect E_M, the mobilization efficiency, by changing the residual oil saturation term. In addition to these effects, E_M is affected by interphase mass transfer that essentially changes the oil formation volume factor. For instance, CO_2 will dissolve in crude oil and increase the volume of the liquid phase to above the end point relative permeability saturation. Mobilization efficiency is also increased by stripping of oil into the vapor phase. Thus, steam distillation of the crude, which occurs in thermal processes, can improve mobilization efficiency. CO_2 can also vaporize significant amounts of hydrocarbons, and for the Weeks Island pilot, this effect is believed to contribute about 35% of the anticipated EOR (Perry and Guillory 1980).

2.5 SAMPLE CALCULATIONS OF RECOVERY FACTORS

2.5.1 Immiscible Displacement: Polymer Flooding in Flat, Layered Beds.
Given: A 40-acre, inverted five spot with the following characteristics:

h = 25 ft, $\quad$ μ_{poly} = 5 cp, $\quad$ β_{poly} = 1.02 bbl/STB,

S_{oi} = 0.8, $\quad$ μ_o = 5 cp, $\quad$ β_o = 1.15 bbl/STB,

k = 260 md, $\quad$ μ_w = 0.5 cp, $\quad$ β_w = 1.02 bbl/STB,

k_σ = 130 md, $\quad\quad\quad\quad\quad\quad$ ϕ = 0.22.

Assume relative permeability curves are the same for polymer and water and no plugging effects or mixing exist. Determine, for both a waterflood and a polymer flood, the following:

WATER. (1) Draw the appropriate fractional flow curve for waterflooding and polymer flooding.

S_w	k_{ro}	k_{rw}	f_w
.2	1.0	0	0
.25	.60	.03	.3333
.3	.38	.05	.5682
.35	.25	.08	.7619
.4	.17	.13	.8844
.45	.10	.16	.9412
.5	.06	.23	.9746
.55	.03	.30	.9901
.6	.01	.35	.9972
.65	0	.45	1.0

S_{wbt} = .26 $\quad\quad$ S_{obt} = .74

$\overline{S}_{wbt}$ = .372 $\quad\quad$ $\overline{S}_{obt}$ = .628

(2) Find E_D as a function of pore volumes injected, and abandon at a WOR of 100. What is S_{orp} equal to?

	S_w	f_w	df_w/dS_w	$\overline{S}_w$	V_{pi}	$\overline{S}_o$	E_D
breakthrough	.26			.372	.16	.628	.382
	.3	.568	3.833	.412	.261	.588	.472
	.35	.762	3.167	.425	.316	.575	.500
	.4	.889	1.5	.474	.667	.526	.609
	.45	.941	.769	.528	1.3	.472	.729
	.5	.972	.467	.560	2.14	.440	.80
	.55	.990	.231	.593	4.33	.407	.874
	.6						
	.65		0				1.0

$$WOR = 100$$

$$f_w = \frac{100\ STB_{H_2O}}{100\ STB_{H_2O} + 1\ STB_{oil}} = \frac{100(1.02)}{100(1.02) + 1(1.15)} = .989$$

f_w	df_w/dS_w	V_{pi}
.989	.222	4.5 ← economic limit $E_D = .88$

(3) Determine the mobility ratio.

k_{ro} at $S_{oi} = .8$	$k_{ro} = 1$	$\mu_o = 5$ cp
k_{rw} at $S_w = .372$	$k_{rw} = .1$	$\mu_w = .5$ cp

$$M = \frac{\lambda_{ing}}{\lambda_{ed}} = \frac{k_w \mu_o}{\mu_w k_o} = \frac{(.1)(5)}{(.5)(1)} = 1$$

(4) Find areal sweep at breakthrough.

$$E_{Abt} = .68\ V_{pibt} = E_{Abt}\ (\overline{S}_{wbt} - S_{wi}) = .68(.372 - .2) = .117$$

(5) Find E_A as a function of pore volumes injected starting at breakthrough.

$$E_A \cong E_{Abt} + .63117\ \log_{10}\left(\frac{V_{pi}}{V_{Pbt}}\right)$$

V_{pi}	E_A
.117	.680
.15	.748
.20	.827
.25	.888
.30	.938
.35	.980
.375	.999
.40	1.0

(6) For layered flow and the permeabilities given, what is E_V at WOR = 100?

$$V_{DP} = .5 \qquad E_V = .96$$

(7) Estimate the mobilization factor for the flood (E_M).

$$E_M = \frac{\dfrac{S_{oi}}{B_{oi}} - \dfrac{S_{orp}}{B_{of}}}{\dfrac{S_{oi}}{B_{oi}}}$$

$$\frac{\dfrac{0.8}{1.15} - \dfrac{0.35}{1.15}}{\dfrac{0.8}{1.15}} = 0.5625$$

POLYMER

(1)

S_P	k_{ro}	k_{rp}	f_p
.2	1.0	0	0
.25	.60	.03	.0476
.3	.38	.05	.1160
.35	.25	.08	.2424
.4	.17	.13	.4333
.45	.10	.16	.6154
.5	.06	.23	.7930
.55	.03	.30	.9091
.6	.01	.35	.9722
.65	0	.45	1

$$S_{pbt} = .513 \qquad S_{obt} = .487$$

$$\overline{S_{pbt}} = .575 \qquad \overline{S_{obt}} = .425$$

(2)

S_p	f_p	df_p/dS_p	$\overline{S}_p$	V_{pi}	$\overline{S}_o$	E_D
.513			.575	.375	.425	.833
.53	.87	2.0	.595	.500	.405	.878
.55	.91	1.5	.610	.667	.390	.911
.57	.94	1.182	.622	.846	.378	.939
.60	.97	0.778	.641	1.29	.359	.980
.65	1	0				1.0

Economic limit $f_P = .989$

f_P	df_p/dS_p	V_{pi}
.989	.65	1.54 $\leftarrow E_D = .984$

(3) k_{ro} at $S_{oi} = .8$ $\qquad k_{ro} = 1 \qquad \mu_o = 5$ cp
k_{rp} at $S_P = .575$ $\qquad k_{rp} = .35 \qquad \mu_p = 5$ cp

$$M = \frac{\lambda_{ing}}{\lambda} = \frac{k_{rp}\mu_o}{\mu_p k_{ro}} = \frac{(.35)(5)}{(5)(1)} = .35$$

(4) $E_{Abt} = .85 \qquad V_{pibt} = E_{Abt} \qquad (\overline{S_{pbt}} - S_{pi}) = .85 \qquad (.575 - .2) = .319$

(5) $E_A \cong E_{Abt} + .63117 \log_{10} \left(\dfrac{V_{pi}}{V_{Pbt}} \right)$

V_{pi}	E_A
.319	.850
.35	.875
.40	.912
.45	.944
.50	.973
.55	.999
.60	1.0

(6) $V_{DP} = .5$ $E_V = .97$

(7) $E_M = \dfrac{\dfrac{0.8}{1.15} - \dfrac{0.35}{1.15}}{\dfrac{0.8}{1.15}} = 0.5625$

2.5.2   Miscible Displacement: Carbon Dioxide Flooding in a Dipping, Homogeneous Bed. The following are given: 600-ft line drive with 10 ft of thickness and 300 ft between injectors:

S_{gi}	$= 0,$	μ_{CO2}	$= 0.05$ cp,	$S_{or,\ wf}$	$= 0.25,$
S_{oi}	$= 0.75,$	μ_o	$= 5$ cp,	α	$= 45°,$
B_{oi}	$= 1.20$ bbl/STB,	μ_w	$= 0.50$ cp,	$GOR_{aband.}$	$= 30 \dfrac{\text{MSCF}}{\text{STB}},$
B_{of}	$= 1.40$ bbl/STB,	ρ_{CO2}	$= 0.65$ gm/cc,	k	$= 3$ darcys,
B_w	$= 1.02$ bbl/STB,	ρ_o	$= 0.90$ gm/cc,	S_{wc}	$= 0.25.$
B_{gCO_2}	$= 5.154 \cdot 10^{-4}$ bbl/SCF,	$\sigma_{o\ -\ CO2}$	$= 0$ dynes/cm,		

(1) The critical velocity for displacement stability (ft/day) is what?

$v_c = \dfrac{\rho_o - \rho_g}{\mu_o - \mu_g}(k\ g \sin \alpha)$

$= \dfrac{.90 - .65}{5 - .05} (3) \left(\dfrac{980}{1.0133 \cdot 10^6} \right) \sin 45°$

$= 1.036 \cdot 10^{-4}$ cm/sec $\cdot \dfrac{1 \text{ in}}{2.54 \text{ cm}} \cdot \dfrac{1 \text{ ft}}{12 \text{ in}} \cdot \dfrac{3,600 \text{ sec}}{1 \text{ hr}} \cdot \dfrac{24 \text{ hr}}{\text{day}}$

$= 2.937 \cdot 10^{-1}$ ft/day

(2) Estimate how many SCF, per well, of CO_2 can be injected daily to maintain this rate.

$$v_c = \frac{Q}{A}$$

$$Q = v_c A = (2.937 \cdot 10^{-1} \text{ ft/day})(10 \text{ ft} \cdot 300 \text{ ft})$$

$$= (8.811 \cdot 10^2 \text{ft}^3)\left(\frac{1}{5.615}\frac{\text{bbl}}{\text{ft}^3}\right)$$

$$= 1.569 \cdot 10^2 \text{ bbl/day} \cdot \frac{1}{5.154 \cdot 10^{-4} \text{ bbl/SCF}} = 3.045 \cdot 10^5 \frac{\text{SCF}}{\text{day}}$$

(3) Given that injectors and producers are 600 ft apart, does this rate sound reasonable? Yes, typical flood rates are between 0.1 and 1 ft/day.

(4) Construct a fractional flow curve for gas displacing oil using Koval's E.

$$f_g = \frac{E S_g}{1 + S_g(E - 1)} \qquad E = \left[.78 + .22\left(\frac{\mu_o}{\mu_s}\right)^{1/4}\right]^4$$

$$f_g = \frac{4.74 S_g}{1 + S_g(4.74 - 1)} \qquad E = \left[.78 + .22\left(\frac{5}{.05}\right)^{1/4}\right]^4$$

$$E = 4.74$$

S_g	f_g
.01	.045
.05	.20
.1	.3449
.2	.542
.3	.670
.4	.760
.5	.826
.6	.877
.65	.898
.7	.917
.8	.950
.9	.977
.95	.989

(5) Calculate E_D as a function of the pore volumes injected. Label breakthrough and economic limit (i.e., abandonment) points.

$$E_D = \frac{2(EV_{pi})^{1/2} - 1 - V_{pi}}{E - 1} = \frac{2(4.74 V_{pi})^{1/2} - 1 - V_{pi}}{3.74}$$

at breakthrough $V_{pi} = 1/E = 1/4.74 = .211 \rightarrow E_D = .211$

V_{pi}	E_D
.211	.211
.3	.290
.4	.362
.5	.422
.6	.474
.7	.520
.8	.560
1.0	.630
1.2	.687
1.4	.736
2.0	.844
2.5	.905
2.75	.928
3.0	.947
3.75	.975
4.0	.992
4.25	.996

$$GOR_{aband.} = \frac{30 \cdot 10^3 SCF}{STB}$$

$$f_{g\,aband.} = \frac{30 \cdot 10^3 \ SCF(5.154 \cdot 10^{-4} \ bbl/SCF)}{30 \cdot 10^3 \ SCF(5.154 \cdot 10^{-4} \ bbl/SCF) + 1 \ STB \ (1.4 \ bbl/STB)}$$

$$= .917$$

$$V_{pi} \text{ (economic limit)} = \frac{E}{[E - f_g(E - 1)]^2}$$

$$= \frac{4.74}{[4.74 - .917(3.74)]^2}$$

$$= 2.76 \ PV$$

(6) Using figure 2.12, what would you estimate the final efficiency (vertical) to be if we flood at the critical rate?

V_{pi} shut down $= 2.76 \rightarrow E_V = 1$

(7) Assuming we inject at the critical rate, what is the capillary number for this flood? If flood is immiscible, $\sigma \simeq 30$ dyne/cm, what is the capillary number?

$N_{cap} = \infty$ for miscible flood

$$N_{cap} = 3.527 \cdot 10^{-4} \frac{V\mu}{\sigma} = \frac{3.527 \cdot 10^{-4}(2.937 \cdot 10^{-1})(.0005)}{30}$$

$$= 1.72 \cdot 10^{-9}$$

(8) What is E_M for a totally miscible displacement?

$$E_M = \frac{\dfrac{S_{oi}}{B_{oi}} - \dfrac{S_{orp}}{B_{of}}}{\dfrac{S_{oi}}{B_{oi}}} = \frac{\dfrac{.75}{1.2} - \dfrac{0}{1.4}}{\dfrac{.75}{1.2}} = 1$$

(9) Determine the mobility ratio assuming miscibility.

$$M = \frac{\mu_{ed}}{\mu_{ing}} = \frac{5}{.05} = 100$$

(10) Given the following gas-water relative permeabilities, determine the proper injection gas-water ratio (SCF/STB) to reduce the displacing phase's mobility:

S_g	k_{rg}	k_{rw}
.05	0	.90
.10	.01	.75
.20	.10	.57
.30	.21	.40
.40	.32	.29
.50	.45	.20
.60	.62	.11
.70	.81	.03
.75	.93	0

$$V_g = \frac{k_{rg}}{\mu_g S_g} \qquad V_w = \frac{k_{rw}}{\mu_w(S_w - S_{wc})}$$

S_w	S_g	V_g	V_w
.9	.1	2	2.31
.8	.2	10	2.07
.7	.3	14	1.78
.6	.4	16	1.66
.5	.5	18	1.6
.4	.6	20.67	1.47
.3	.7	23.14	1.2
.25	.75	24.8	∞

$$V_g = V_w \text{ at } S_w \cong .9$$

$$\text{Gas-water injection ratio} = \frac{.1(\text{bbl})}{(.9 - .25)(\text{bbl})} = \frac{S_g}{S_w - S_{wc}}$$

$$= \frac{(.1)(\text{bbl})/5.154 \cdot 10^{-4} \ (\text{bbl/SCF})}{(.65)(\text{bbl})/1.02 \ (\text{bbl/STB})}$$

$$= \frac{304 \ \text{SCF CO}_2}{1 \ \text{STB H}_2\text{O}}$$

REFERENCES

Bilhartz, H. L. Jr., Charlson, G. S., and Stalkup, F. I.: "A Method for Projecting Full-Scale Performance of CO_2 Flooding in the Willard Unit," paper SPE 7051 presented at the SPE 5th Symposium on Improved Methods for Oil Recovery, Tulsa, April 16–19, 1978.

Blackwell, R. J., Terry, W. M., Rayne, J. R., and Lindley, D. C.: "Recovery of Oil by Displacements with Water-Solvent Mixtures," *Trans., AIME* (1960) **219**, 293–300.

Buckley, S. E., and Leverett, M. C.: "Mechanism of Fluid Displacement in Sands," *Trans., AIME* (1942) **146**, 107–116.

Caudle, B. H., and Dyes, A. B.: "Improving Miscible Displacement by Gas-Water Injection," *Trans., AIME* (1958) **213**, 281–284.

Caudle, B. H., and Witte, M. D.: "Production Potential Changes During Sweep-Out in a Five-Spot System," *J. Pet. Tech.* (Dec. 1959) 63–65.

Craig, F. F.: *The Reservoir Engineering Aspects of Waterflooding*, Monograph Series, SPE, Dallas (1971) **3**, 29–47.

Craig, F. F., Sanderlin, J. L., Moore, D. W., and Geffen, T. M.: "A Laboratory Study of Gravity Segregation in Frontal Drives," *J. Pet. Tech.* (Oct. 1957) 275–280.

Dumoré, J. M.: "Stability Considerations in Downward Miscible Displacements," *Soc. Pet. Eng. J.* (Dec. 1964) 356–362.

Dykstra, H., and Parsons, R. L.: *Secondary Recovery of Oil in the United States*, API, Washington, D.C. (1950).

Fayers, F. J., Hawes, R. I., and Mathews, J. D.: "Some Aspects of the Potential Application of Surfactants or CO_2 as EOR Processes in North Sea Reservoirs," *J. Pet. Tech.* (Sept. 1981) 1617–1627.

Gardner, J. W., and Ypma, J. G. J.: "An Investigation of Phase Behavior-Macroscopic Bypassing Interaction in CO_2 Flooding," paper SPE/DOE 10686 presented at the SPE/DOE Third Joint Symposium on Enhanced Oil Recovery, Tulsa, April 4–7, 1982.

Gupta, S. P., and Trushenski, S. P.: "Micellar Flooding—Compositional Effects on Oil Displacement," *Soc. Pet. Eng. J.* (April 1979) 116–128.

Hawthorne, R. G.: "Two-Phase Flow in Two-Dimensional Systems—Effects of Rate, Viscosity and Density on Fluid Distribution in Porous Media," *Trans., AIME* (1960) **219**, 81.

Kane, A. V.: "Performance Review of a Large-Scale CO_2-WAG Enhanced Recovery Project, SACROC Unit—Kelly-Snyder Field," *J. Pet. Tech.* (Feb. 1979) 217–231.

Koval, E. J.: "A Method for Predicting the Performance of Unstable Miscible Displacement in Heterogeneous Media," *Soc. Pet. Eng. J.* (June 1963) 145–154.

Lantz, R. B.: "Rigorous Calculation of Miscible Displacement Using Immiscible Reservoir Simulators," *Soc. Pet. Eng. J.* (June 1970) 192.

Perry G. E., and Guillory, A. J.: "Weeks Island S Sand Reservoir B Gravity Stable Miscible CO_2 Displacement," DOE Contract EF(77–C–05–5232), (1980).

Raimondi, P., and Torcaso, M. A.: "Distribution of the Oil Phase Obtained upon Imbibition of Water," *Soc. Pet. Eng. J.* (Mar. 1964) 49–55.

Shelton, J. L., and Schneider, F. N.: "The Effects of Water Injection on Miscible Flooding Methods Using Hydrocarbons and Carbon Dioxide," *Soc. Pet. Eng. J.* (June 1975) 217–226.

Stahl, C. D.: *Waterflood Theory,* personal notes (1960) 142.

Stalkup, F. I.: "Displacement of Oil by Solvent at High Water Saturation," *Soc. Pet. Eng. J.* (Dec. 1970) 337–348.

Stalkup, F. I.: "Carbon Dioxide in Carbonates and Sandstones," paper presented at the Symposium on Technology of EOR in the Year 2000, Williamsburg, Va., June 28–29, 1980, 157–179.

Stalkup, F. I., Jr.: *Miscible Displacement,* Monograph Series, SPE, Dallas (1983) **8,** 137–158.

Stegemeier, G. L.: *Improved Oil Recovery by Surfactant and Polymer Flooding,* Academic Press, New York, 1977.

Stiles, W. E.: "Use of Permeability Distribution in Waterflood Calculations," *Trans., AIME* (1949) **186,** 9–13.

Tiffin, D. L., and Yellig, W. F.: "Effects of Mobile Water on Multiple Contact Miscible Gas Displacements," paper SPE/DOE 10687 presented at the 3rd Joint Symposium on EOR, Tulsa, April 4–7, 1982.

Todd, M. R., and Longstaff, W. J.: "The Development, Testing, and Application of a Numerical Simulator for Predicting Miscible Flood Performance," *J. Pet. Tech.* (July 1972) 874–882.

Warner, H. R.: "An Evaluation of Miscible CO_2 Flooding in Water-Flooded Sandstone Reservoirs," *J. Pet. Tech.* (Oct. 1977) 1339–1347.

Welge, H. J.: "A Simplified Method for Computing Oil Recovery by Gas or Water Drive," *Trans., AIME* (1952) **195,** 91–98.

Youngren, G. K., and Charlson, G. S.: "History Match Analysis of the Little Creek CO_2 Pilot Test," *J. Pet. Tech.* (Nov. 1980) 2042–2052.

3 CARBON DIOXIDE–CRUDE OIL INTERACTION: ITS EFFECT ON RECOVERY

Predictions from any mathematical model are only as accurate as the analytical equations that are used to represent the displacement process. These equations, such as those following in chapter 4, are generally based on many assumptions concerning the displacement mechanisms present.

Equally important, once a reservoir simulator has been developed, is the physical description of the reservoir and the fluids it contains. In predicting future pressure-production behavior, input data such as reservoir size, pay thickness, depth, porosity, permeability, fluid properties, rock properties, and saturations are based on limited well test, log, and core data from a minute portion of the reservoir. Since the true reservoir description is never known, a careful screening of laboratory and field data is necessary to insure meaningful simulation results. Only after this is done can multiple simulations be made to determine which variables have a major effect on reservoir performance. By altering these variables a family of potential pressure-production scenarios can be examined, within which future reservoir performance should fall.

Mathematical representations of the rock and fluid properties in situ must be derived. Data such as gas viscosity as a function of pressure and temperature, the solubility of CO_2 in crude oil, and so forth must be available to the reservoir modeler. The purpose of this chapter, then, is to present the types of information necessary to predict results of CO_2 injection adequately and to show how these data are usually obtained.

3.1 PURE CARBON DIOXIDE PROPERTIES

3.1.1 **Gas Compressibility, Z.** The perfect gas law, which expresses the relationship between pressure, temperature, and volume of an ideal gas, is a combination of Boyle's, Charles's, and Avogadro's Laws:

$$PV = nRT. \tag{3.1}$$

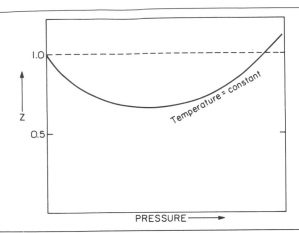

Figure 3.1 Typical plot of the gas supercompressibility *(z)* factor as a function of pressure at a constant temperature.

Since no gases react ideally at all temperatures and pressures, two approaches were brought about to model the behavior of an actual gas: Van der Waals's Equation and the Imperfect Gas Law. Van der Waals's Equation for n moles of a single, pure gas is written

$$\left(P + \frac{n^2 a}{V^2}\right)(V - nb) = nRT, \tag{3.2}$$

where a and b are constants whose values are different for each gas (Burcik 1956). The quantity a/V^2 accounts for the attractive forces between the molecules. It is added to the pressure because the actual pressure would need to be larger to produce the same volume than if no attraction existed. The constant b represents the volume of the molecules, and it is substracted from V since the actual volume of space available to the gas is less than the overall volume of the gas.

Van der Waals's Equation is unwieldy for most engineering calculations, and consequently, a second method, the Imperfect Gas Law, was developed. This law in its general form can be written

$$PV = ZnRT, \tag{3.3}$$

where Z is known as the compressibility, or deviation, factor. It is an empirical factor, determined experimentally, that makes equation (3.3) true at a particular temperature and pressure. At a given temperature the Z factor plotted as a function of pressure usually takes the form shown in figure 3.1.

The deviation of CO_2 from ideal gas behavior, Z, is given as a function of pressure and temperature in figure 3.2 (Reamer et al. 1944).

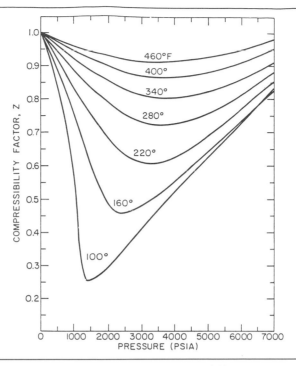

Figure 3.2 Compressibility factors for CO_2 (Reamer et al. 1944).

3.1.2 **Gas Formation Volume Factor, B_g.** Gas volume factors relate the volume of gas in the reservoir to the volume on the surface—that is, at standard conditions, P_{sc} and T_{sc}. Assuming $Z_{sc} = 1$ and setting $V_{sc} = 1$ SCF, then B_g is defined as

$$B_g = \frac{P_{sc} Z\, T}{T_{sc} P} \,.$$ (3.4)

If P_{sc} is 14.7 psia and $T_{sc} = 520°R$, for a given P and T,

$$B_g = 0.02827 \frac{ZT}{P} \quad \text{(cu ft/SCF)}$$ (3.5a)

$$= 0.00504 \frac{ZT}{P} \quad \text{(bbl/SCF)}$$ (3.5b)

$$= 35.37 \frac{P}{ZT} \quad \text{(SCF/cu ft)}$$ (3.5c)

$$= 198.6 \frac{P}{ZT} \quad \text{(SCF/bbl).}$$ (3.5d)

3.1.3

Density, ρ. CO_2 exists normally as a gas and possibly as a liquid at normal reservoir temperature and pressure. Having a critical temperature of 87.8°F precludes most reservoirs from liquid CO_2 displacement. CO_2 is, however, a relatively dense gas with a 50% greater density than air at atmospheric conditions and a much lower compressibility factor at typical reservoir conditions. In figure 3.3, Stalkup (1983) shows, in more detail, the solid-liquid-vapor equilibrium relationship for CO_2 below its critical temperature.

Above the critical temperature, CO_2 behaves as a vapor whose density increases as pressure increases. Figure 3.4 shows that fluid density is a continuous function of pressure at temperatures above critical conditions but that abrupt discontinuities will appear at pressures below the critical temperature. Also note that near the critical region, CO_2 densities approach those of the displaced oil and may, in fact, be heavier than the resident hydrocarbons.

In addition, gas density (lb/ft^3) at low pressures may be calculated by

$$\rho = \frac{PM}{ZRT},\tag{3.6}$$

where

P = pressure, psia,
M = molecular weight (CO_2 = 44.01),
Z = gas supercompressibility,
T = temperature, °R,
R = universal gas constant, 10.73.

3.1.4

Viscosity, μ_g. The viscosity of CO_2 is a strong function of pressure and temperature. This effect is best shown in figure 3.5 (Goodrich 1980). Note that as pressure increases at a constant reservoir temperature, gas viscosity increases considerably. Curves of this type are also available for most gas systems. In comparing the viscosity of CO_2 with other pure components, low pressure viscosities (measured at reservoir temperature and 1 atm) for several common gases are shown in figure 3.6 (Carr et al. 1954).

3.1.5

Enthalpy, H. The enthalpy, or heat content of a material, is a thermodynamic quantity equal to the sum of the internal energy of a system plus the product of the pressure-volume work done on the system. By definition,

$$H = U + pV\tag{3.7}$$

Thus, the change in enthalpy is given by

$$\Delta H = \Delta U + \Delta pV.\tag{3.8}$$

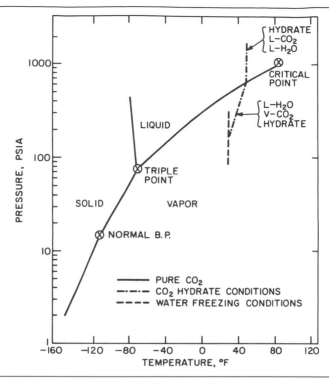

Figure 3.3 Phase behavior of CO_2 and CO_2-water mixtures (Stalkup 1983; Unruh and Katz 1949; Robinson and Mehta 1971) ©SPE-AIME.

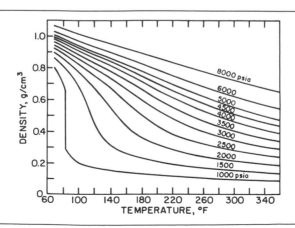

Figure 3.4 Density of CO_2 as a function of pressure and temperature (Holm and Josendal 1982; Vukalovich and Altunin 1968) ©SPE-AIME.

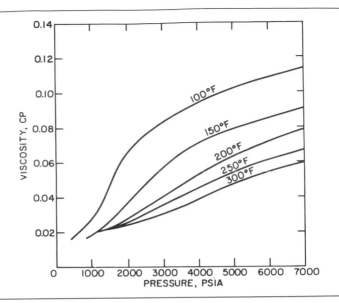

Figure 3.5 Viscosity of CO_2 as a function of pressure (Goodrich 1980).

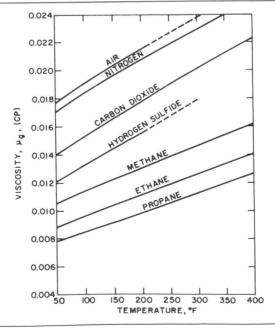

Figure 3.6 Viscosity of several common gases at atmospheric pressure and various temperatures (Carr et al. 1954) ©SPE-AIME.

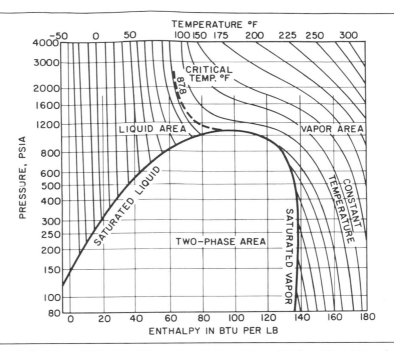

Figure 3.7 Enthalpy of CO_2 from *Thermodynamic Properties and Reduced Correlations for Gases* by Lawrence N. Canjar and Francis S. Manning. Copyright ©1967 by Gulf Publishing Company, Houston, Texas. All rights reserved. Used with permission.

The change in enthalpy in a system is normally associated with the design of horsepower requirements for compressors under different loading factors. In CO_2 injection, enthalpy changes must be accounted for in determining tophole or bottomhole pressures where phase changes occur down the wellbore. The effect of pressure and temperature on the heat content, H, of a CO_2 system is given in figure 3.7 (Canjar and Manning 1967). A more complete discussion of enthalpy effects on CO_2-wellbore phenomena is forthcoming in chapter 4.

3.2   **CARBON DIOXIDE AS A DISPLACEMENT FLUID.** The use of CO_2 to increase oil recovery is not a new idea. In 1952, Whorton et al. received the first patent for oil recovery using CO_2. These early investigators considered using CO_2 as a solvent for crude oil or as a carbonated waterflood. Since then, many injection schemes using CO_2 liquid and gas have been suggested, including

continuous CO_2 gas injection,
carbonated water injection (ORCO),
CO_2 gas or liquid slug followed by water,

CO_2 gas or liquid slug followed by alternate water and CO_2 gas injection (WAG), and simultaneous injection of CO_2 gas and water.

Figure 3.3 showed the phase behavior of pure CO_2. Note that CO_2 has a critical temperature of 30°C (87°F). While we conclude that liquid or gaseous CO_2 may be injected at the surface (which we discuss in the next chapter), this critical temperature usually precludes most reservoirs from having liquid CO_2 present at the sandface.

Continuous CO_2 injection is an important process to identify displacement mechanisms but is not likely to be economic in practice unless significant recycling of gas is employed. Inherent in all gas injection processes is the lack of mobility and gravity control (areal and vertical sweep) necessary to sweep significant portions of the reservoir. Therefore, the replacement of high cost CO_2 by a cheaper chase fluid such as water for horizontal displacements or nitrogen for gravity stable floods appears economically attractive.

For carbonated water injection, as shown in figure 3.8A, CO_2 diffuses out of the injected water-CO_2 mixture when in contact with the reservoir oil. The diffusion is slow when compared with the injection of a pure CO_2 slug; thus, the chances of obtaining an effective CO_2 concentration at the displacement front are increased when material is injected as a pure slug. Holm (1963) confirmed this premise, and his work is summarized in figure 3.9. In this study, equal pore volumes of CO_2 were injected as a CO_2 slug and as carbonated water while displacing a 5-cp crude. After 2 PV had been injected (400 SCF/bbl of oil in place), the CO_2 slug process recovered 25 to 35% more oil than the carbonated water process.

In figure 3.8B, the CO_2 slug process, CO_2 is followed by continuous water injection to drive the slug through the reservoir. The trailing water immiscibly displaces CO_2, leaving a residual CO_2 saturation in the reservoir. A variation of this process, figure 3.8C, the WAG process, alternates small volumes (5% PV or less) of CO_2 and water until the desired volume of CO_2 (usually 15 to 20% PV) has been injected. The objective is to reduce CO_2 mobility and to promote more uniform vertical entry into the reservoir, thus yielding greater volumetric conformance. After several alternating slugs of CO_2 and water have been injected, continuous injection of water begins.

Another proposed process includes the simultaneous injection of water and CO_2 through dual injection systems (fig. 3.8D). Water is injected in the top of the pay zone and allowed to segregate downward as it flows through the reservoir while CO_2 is injected in the bottom and allowed to rise. However, improved recovery may be more than offset by high operating and completion costs.

An additional mobility control alternative, chasing the CO_2 slug with stabilized foams, is now being considered. In theory, an injected foam bank should reduce the mobility of the gas phase and, hence, improve areal sweep efficiency. The major concern in foam applications, however, is creating a long-term, stable foam at an economically attractive price.

The introduction of mobile water in a WAG process may trap oil by water shielding and CO_2 bypassing as well as introduce potential gravity segregation problems. Work by Tiffin and Yellig (1982) substantiated the fact that *injected* water in tertiary floods with water-wet cores resulted in significantly (20 percentiles) lower oil recovery than continuous CO_2 injection. This was not the case for oil-wet cores, inferring that wetta-

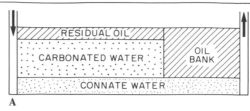

Figure 3.8A Schematic of carbonated waterflooding process (ORCO).

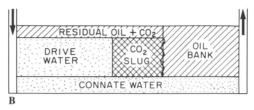

Figure 3.8B Schematic of CO$_2$ slug and water drive process.

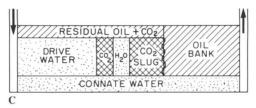

Figure 3.8C Schematic of water-alternating CO$_2$ gas (WAG) process.

Figure 3.8D Schematic of simultaneous water-CO$_2$ injection process.

bility may be a key factor in the performance of a WAG process with respect to oil trapping.

Stalkup (1978) has shown the importance of mobile water on gravity segregation. Using a miscible-flood reservoir simulator for a homogeneous sandstone reservoir that had been waterflooded prior to being CO$_2$ flooded, a 25% hydrocarbon PV slug of CO$_2$ was followed by continuous water injection. As the ratio of vertical/horizontal permeability varied from 0 to 0.1, segregation reduced oil recovery by a factor of about 2.5.

Injecting water to achieve better areal sweep efficiency not only traps a residual CO$_2$ saturation and may cause serious problems such as gravity segregation and oil trapping,

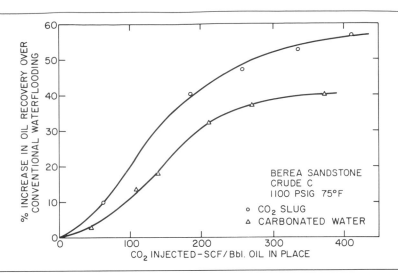

Figure 3.9 Oil recovery using carbonated water and CO_2 slugs (Holm 1963).

but also, CO_2 mixed with water forms carbonic acid, which is highly corrosive. Special metal alloys and coatings for certain facilities are needed. Also, dual injection systems are needed when an alternating injection scheme (WAG) of CO_2 and water is employed.

However, even with these drawbacks, when areal sweep forces dominate design, a WAG scheme still appears to be the most economically feasible CO_2 process except in low permeability sands where alternating injection may seriously reduce injection rate due to the much lower mobility of water. The superiority of WAG injection over continuous or single slug injection was shown in two simulation studies by Warner (1977) and Fayers et al. (1981). Warner's results show continuous CO_2 injection to recover 20% of the potential tertiary oil. Slug CO_2 injection recovered 25%, while the WAG process produced 38% of the postwaterflood oil in place. Simultaneous injection recovered 47% of the oil but still poses serious operating questions.

The earlier discussion was concerned with altering the mobility of CO_2 displacing banks in order to increase *areal* sweep. In reservoirs with a steep dip, the key design parameter may be instead to maximize *vertical* sweep efficiency. A gravity stabilized flood can be designed such that by displacing oil downdip below a critical rate, gravity segregation will prevent formation of viscous fingers and will create a high displacement efficiency (see chapter 2). Economic displacement rates can only be attained in high permeability, steep beds where the density difference between CO_2 and oil is significant. CO_2 may be cut with methane, ethane, and so forth to increase this density difference since the difference between CO_2 and crude oil may be small. To reduce overall costs, CO_2 is normally chased by N_2 or flue gas where density rather than mobility control is the key sweep efficiency variable. However, capital expenses such as pipelines and compressor's are not reduced, while the risk of the chase gas diluting the CO_2 slug is introduced.

Regardless of how CO_2 is injected into the reservoir, oil displacement by CO_2 injec-

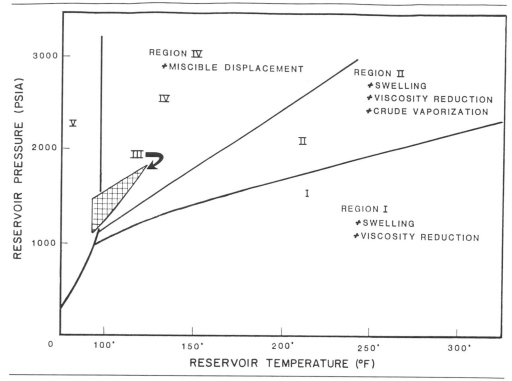

Figure 3.10 The effect of reservoir temperature and pressure on CO₂ injection displacement mechanisms.

tion relies on a number of mechanisms related to the phase behavior of CO_2–crude oil mixtures. Strongly dependent on reservoir temperature, pressure, and crude oil composition, the dominant displacement characteristics for a given CO_2 displacement fall into one of the five regions as shown in figure 3.10:

Region I: Low pressure applications
Region II: Intermediate pressure, high temperature applications
Region III: Intermediate pressure, low temperature applications
Region IV: High pressure applications
Region V: High pressure, low temperature (liquid) applications

It is important to note that the lines that divide region from region are generalizations that will vary from oil to oil. Heavier oils will shift these divisions upward. Since CO_2 exhibits a critical temperature of 87°F, most reservoirs are excluded from having liquid CO_2 present at the sandface. This being the case, little research has been done utilizing Region V liquid CO_2 applications. However, this may change as the limits of reservoir applications of CO_2 are expanded.

3.3   **LOW PRESSURE APPLICATIONS.** At reservoir pressures below 1,000 psia (Region I), the major effects of CO_2 injection on oil recovery appear due to the solubility of CO_2 in the crude oil. Besides an increase in reservoir pressure, the introduction of CO_2 in the reservoir

> swells the oil,
> reduces oil viscosity significantly,
> contributes to internal solution gas drive, and
> increases injectivity.

This low pressure application of CO_2 injection is the type of process that may be best applied in shallow, viscous oil fields where water or thermal methods are inefficient. An application of such a process is discussed by Reid and Robinson (1981) concerning Phillips's Lick Creek immiscible CO_2 project. Performance from this project is estimated to recover over 3 million barrels of incremental oil (17° API) in 15 years while injecting 8.5 billion SCF. A list of additional heavy oil, low pressure field tests is given in chapter 6.

Swelling of oil: Already mentioned is the fact that CO_2 is highly soluble in hydrocarbon oils. Depending upon the saturation pressure, reservoir temperature, and composition of the crude oil, figure 3.11 shows that for a 17° API crude oil, over 700 SCF of CO_2 will dissolve in 1 barrel of oil, yielding a 10 to 30% increase in volume (fig. 3.12) (Miller and Jones 1981).

This swelling is important for two reasons. One, the residual oil left in the reservoir after flooding is inversely proportional to the swelling factor; that is, the greater the swelling, the less stock tank oil abandoned in the reservoir. Second, swollen oil droplets will force water out of pore spaces, creating a drainage rather than imbibition process for water-wet systems. Drainage oil relative permeability curves are higher than their imbibition counterparts, creating a more favorable oil flow environment at any given saturation conditions.

Viscosity reduction: As CO_2 gas saturates a crude oil, a large reduction in the viscosity of that oil occurs. This reduction, like heating of crudes in thermal recovery, can yield viscosities one-tenth to one one-hundredth of the original viscosity. An example of this reduction is shown in figure 3.13 (Miller and Jones 1981). Note that a larger percentage reduction occurs in the viscosity of more viscous crudes.

For example, at a saturation pressure of 1,000 psia, the more viscous crude reduces its viscosity from 120 cp to 10 cp, a twelvefold decrease, while the 18-cp crude undergoes a threefold decrease to 6 cp. Thus, the viscosity reduction and its effect on mobility ratio is more significant in medium and heavy oils and not as large in low viscosity oils. This effect for higher viscosity crudes (>20 cp) has been documented in core floods by Holm (1963) and deNevers (1966).

Solution gas drive: Just as CO_2 goes into solution with an increase in reservoir pressure, after termination of the injection phase of a flood, gas will come out of solution and continue to drive oil into the wellbore. This mechanism of blowdown recovery is similar to solution gas drive during the normal production depletion of an oil field. Holm and Josendal (1974) have shown that up to 18.6% of the oil in place can be recovered by CO_2 solution gas drive, while Wang and Locke (1980) found blowdown recoveries

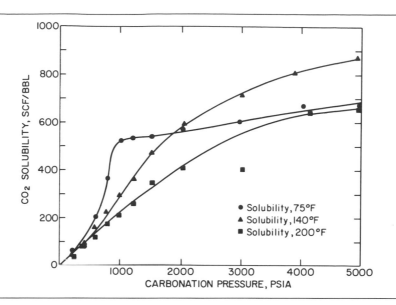

Figure 3.11 Solubility of CO_2 in Wilmington Oil (17° API Gravity) at 75°F, 140°F and 200°F (Miller and Jones 1981) ©SPE-AIME.

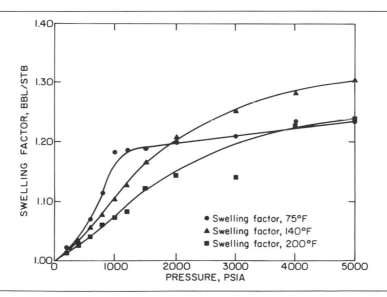

Figure 3.12 Swelling factors of CO_2-saturated Wilmington Crude Oil (17° API Gravity) at 75°F, 140°F and 200°F (Miller and Jones 1981) ©SPE-AIME.

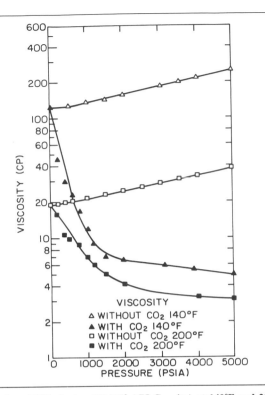

Figure 3.13 Viscosity of Wilmington Oil (17° API Gravity) at 140°F and 200°F (Miller and Jones 1981) ©SPE-AIME.

from 4.73 to 8.55% for a Drake mineral oil system. In high viscosity crudes, Klins and Farouq Ali (1982) found solution gas drive recoveries from 0.3 to 2% of the original oil in place.

Increased injectivity: CO_2-water mixtures are slightly acidic and react accordingly with the formation matrix. In shales, carbonic acid stabilizes clays due to a reduction in ph:

$$CO_2 + H_2O \longrightarrow H_2CO_3. \tag{3.9a}$$

In carbonates, injectivity is improved by partially dissolving the reservoir rock according to the following reactions:

$$H_2CO_3 + CaCO_3 \longrightarrow Ca(HCO_3)_2 \tag{3.9b}$$

and

$$H_2CO_3 + MgCO_3 \longrightarrow Mg(HCO_3)_2. \tag{3.9c}$$

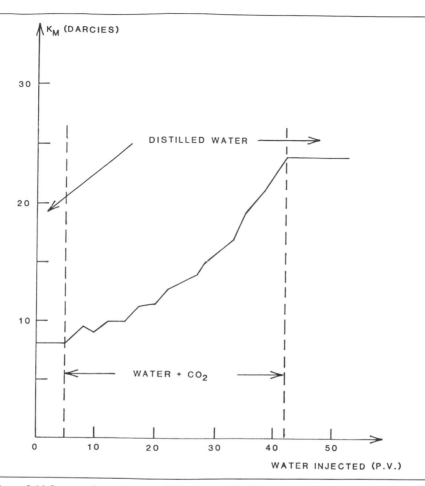

Figure 3.14 Increase in rock permeability due to CO_2 injection (Latil 1980).

The bicarbonates formed are quite soluble in water, which may lead to a permeability increase in carbonate rocks, especially around the wellbore where large volumes of CO_2 and water pass. Figure 3.14 shows the results of flooding a core sample of calcareous rock with carbonated water and the resulting threefold increase in permeability. Additional work has been completed by Ross et al. (1982) on North Sea calcareous sandstones (see also Latil 1980).

There is, however, a possibility that injectivity may be reduced. This dissolution of carbonate materials may free unreacted reservoir fines to flow and later to plug downstream pore channels. Also, plugging due to the precipitation of calcium sulfate or asphaltenes (discussed later in this chapter) may offset any injectivity gains by the reaction of carbonate materials.

3.3.1   A Review of Natural Gas–Crude Oil Properties

GAS SOLUBILITY. The solubility of natural gas in oil must be referred to some basis, and for this purpose, it is customary to use one barrel of stock tank oil. The gas solubility (R) is defined as the number of cubic feet of gas measured at standard conditions that are in solution in one barrel of stock tank oil at reservoir temperature and pressure. A typical gas solubility curve as a function of pressure is shown in figure 3.15 for a saturated crude oil at reservoir temperature. P_o denotes the original reservoir pressure, and R_o is the original value of the gas solubility. As the pressure is reduced, solution gas is liberated and the value of R decreases as shown.

If the pressure is released from a sample of reservoir crude oil, the quantity of gas evolved depends upon the conditions of liberation. There are two basic types of gas liberation: flash and differential. In a flash liberation, the pressure is reduced by a finite amount, and after equilibrium is established, the gas is bled off, keeping the pressure constant. In a differential liberation, the gas evolved is removed continuously from contact with the oil. The liquid is in equilibrium only with the gas being evolved at a given pressure and not with the gas evolved over a finite pressure range. It is apparent that a series of flash liberations with infinitely small pressure reductions approach a differential liberation. The two methods of liberation give different results for R, as shown in figure 3.16, the values of R for flash liberation being higher at a given pressure. It is difficult to say which type of liberation is operative in a reservoir, and in all probability both occur simultaneously.

The best known method of estimating the amount of natural gas dissolved in a given crude oil is that of Standing (1947). An improved correlation was developed by Vazquez and Beggs (1980). Their equation, which follows, was obtained by regression analysis using 5,008 data points:

$$R_s = C_1\gamma_{gs}p^{C_2}\exp\{C_3[\gamma_o/(T + 460)]\} \tag{3.10}$$

Values for the coefficients are as follows:

Coefficient	$\gamma_o \leq 30$	$\gamma_o > 30$
C_1	0.0362	0.0178
C_2	1.0937	1.1870
C_3	25.7240	23.9310

where

R_s = dissolved GOR, SCF/STB,
γ_{gs} = modified gas gravity (air = 1) [see eq. (3.14)],
P = pressure, psia,
γ_o = oil gravity, °API,
T = temperature, °F.

OIL-FORMATION VOLUME FACTOR. If the pressure on a sample of reservoir oil is decreased below its saturation pressure, gas is evolved and the volume of the residual

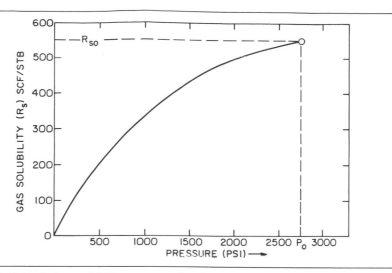

Figure 3.15 Typical gas solubility curve as a function of pressure for a saturated crude oil.

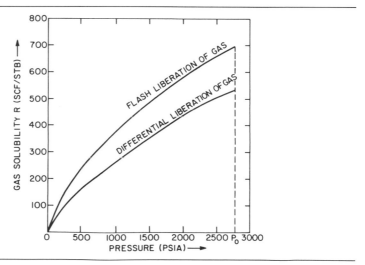

Figure 3.16 Typical plot of gas solubility versus pressure showing the difference obtained by flash and differential liberation.

oil decreases. The shrinkage in oil volume upon liberation of gas may be expressed as a relative volume ratio. Here again, the basis may be either the original or the final oil volume. Consider the process shown in figure 3.17. A sample of reservoir liquid has a volume of V_1 barrels under reservoir conditions of temperature and pressure. When this liquid is brought to stock tank conditions, the volume is reduced to V_2 barrels such that $B_o = V_1/V_2$.

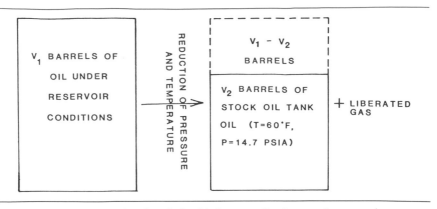

Figure 3.17 Diagram showing the relationship between oil volume under reservoir conditions and under stock tank conditions.

In engineering calculations, the formation volume factor, B_o, is most commonly used to express the change in liquid volume with pressure. B_o is defined as the volume in barrels at reservoir pressure and temperature occupied by one barrel of stock tank oil including the gas in solution at that temperature and pressure. The change in B with pressure for a typical saturated crude oil is shown in figure 3.18. The original reservoir pressure is P_o, and the original value of the formation volume factor is represented by B_o. As the pressure is reduced below P_o, solution gas is evolved, the volume of the oil is reduced, and the value of B is decreased. When the pressure is reduced to atmospheric, the value of B_o is nearly equal to 1. ($B_o = 1$ when the reservoir temperature is 60°F.)

As pointed out in the discussion of solution gas, the value of R depends on the method of gas liberation. It is evident that the value of the formation volume factor is also dependent on the method of gas liberation, as shown in figure 3.19. Since less gas is evolved on differential liberation, the residual oil volume is greater. Consequently, the differential formation volume factor is less than the flash formation volume factor as shown.

Similar to the gas solubility correlation, the best known work on determining the oil formation volume factor was that of Standing (1947). Vazquez and Beggs (1980) developed an oil FVF (formation volume factor) correlation for saturated oil as a function of dissolved gas, temperature, oil gravity, and gas gravity. The following formula represents their results:

$$B_o = 1 + C_1 R_s + C_2(T - 60)(\gamma_o/\gamma_{gs}) + C_3 R_s(T - 60)(\gamma_o/\gamma_{gs}). \tag{3.11}$$

The values for the coefficients depend on oil gravity and are given by the following:

Coefficient	$\gamma_o \leq 30$	$\gamma_o > 30$
C_1	4.677×10^{-4}	4.670×10^{-4}
C_2	1.751×10^{-5}	1.100×10^{-5}
C_3	-1.811×10^{-8}	1.337×10^{-9}

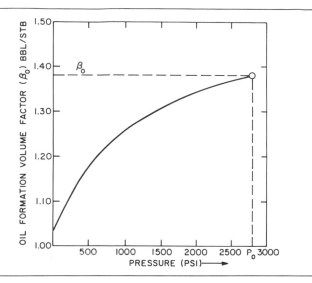

Figure 3.18 Typical plot showing the dependence of the formation volume factor on pressure for a saturated crude oil.

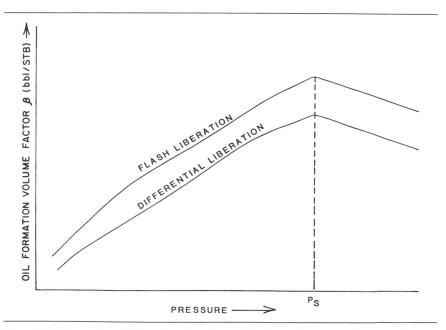

Figure 3.19 Illustrative plot showing dependence of the formation volume factor on the method of gas liberation.

OIL COMPRESSIBILITY. Oil compressibilities are used to calculate the formation volume factor, B_o, and the oil density, ρ_o, above the bubble point. For example, if B_{ob} represents the oil formation volume factor at the bubble point, then:

$$B_o = B_{ob} \exp[c_o(P_b - P)] \tag{3.12a}$$

and

$$\rho_o = \rho_{ob} \exp[c_o(P - P_b)]. \tag{3.12b}$$

In determining the compressibility, c_o, of an undersaturated oil, two correlations have been developed. The first, by Standing (1947), is shown in figure 3.20. The second, a mathematical correlation, was developed by Vazquez and Beggs (1980). In this correlation, c_o was expressed as a function of R_{so}, T, γ_o, γ_{gs}, and P. A total of 4,036 data points were used in a linear regression to obtain

$$c_o = (a_1 + a_2 R_{so} + a_3 T + a_4 \gamma_{gs} + a_5 \gamma_o)/a_6 P, \tag{3.13}$$

where

$a_1 = -1433,$
$a_2 = 5,$
$a_3 = 17.2,$
$a_4 = -1180,$
$a_5 = 12.61,$
$a_6 = 10^5,$

and

c_o = oil compressibility, 1/psi,
R_{so} = dissolved GOR, SCF/STB,
T = reservoir temperature, °F,
γ_{gs} = modified gas gravity (air = 1),
γ_o = oil gravity, °API,
P = oil pressure, psia,

where

$$\gamma_{gs} = \gamma_{gp}[1 + 5.912 \cdot 10^{-5} (\gamma_o) \cdot (T) \log (p/114.7)] \tag{3.14}$$

and

γ_{gs} = gas gravity (air = 1) that would result from separator conditions of 100 psig,
γ_{gp} = gas gravity obtained at separator conditions of p and T,
p = actual separator pressure, psia,
T = actual separator temperature, °F,
γ_o = oil gravity, °API.

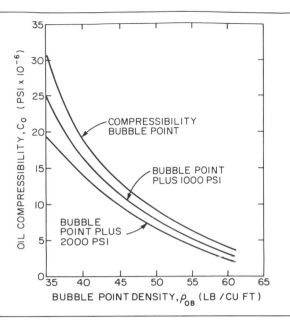

Figure 3.20 Compressibility of reservoir crude oils (Standing 1947).

VISCOSITY OF RESERVOIR OILS. The viscosity of a fluid, which is a measure of its resistance to flow, is defined as the force in dynes on a unit area of either of two horizontal planes a unit distance apart, one of which is fixed while the other moves with unit velocity, the space between the planes being filled with the viscous fluid.

In the case of hydrocarbon liquids certain generalizations regarding viscosity can be made, viz.,

Viscosity decreases with increasing temperature;
Viscosity increases with increasing pressure, provided the only effect of pressure is to compress the liquid;
Viscosity decreases as the gas in solution increases.

For most reservoir liquids the effect of liquid compression increasing viscosity is more than counterbalanced by the effect of solution gas so that oil viscosity decreases with increasing pressure until the saturation pressure is reached. A further increase in pressure will cause an increase in viscosity due to compression of the liquid.

If experimental data for the crude oil viscosity are not available, an estimate can be made in the following manner. Beal (1946) drew the first correlation between oil gravity, reservoir temperature, and dead oil viscosity. His results are shown in figure 3.21. Chew and Connally (1959) extended Beal's work by correlating Beal's dead oil viscosity with natural-gas-saturated crude viscosity. Their correlation is shown in figure 3.22. For viscosities above the bubble point, figure 3.23 by Beal may be used.

If, instead, viscosity as a function of pressure is required, one may use correlations

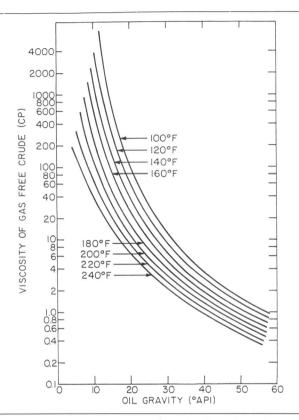

Figure 3.21 Gas-free oil viscosity as a function of crude oil gravity and temperature (Beal 1946) ©SPE-AIME.

developed by Beggs and Robinson (1975) and Vazquez and Beggs (1980). They state that the viscosity of a gas-free crude at temperature T (°F) is given by

$$\mu_{od} = 10^X - 1, \tag{3.15}$$

where

$$X = yT^{-1.163},$$
$$y = 10^Z,$$
$$Z = 3.0324 - 0.02023\gamma_o.$$

Live (gas-saturated) oil viscosity may then be calculated from

$$\mu = A\mu_{od}^B, \tag{3.16}$$

where

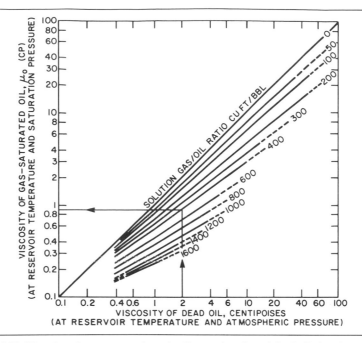

Figure 3.22 Viscosity of gas-saturated crude oils as a function of dead oil viscosity and natural gas solubility (Chew and Connally 1959) ©SPE-AIME.

$$A = 10.715(R_{so} + 100)^{-0.515},$$
$$B = 5.44(R_{so} + 150)^{-0.338},$$

and

R_{so} = dissolved GOR, SCF/STB,
T = temperature, °F,
μ_{od} = viscosity of gas-free oil at T, cp,
γ_o = oil gravity, °API.

3.3.2 **Carbon Dioxide–Crude Oil Systems.** For low pressure CO_2 displacements, there is little or no mass transfer from the oil phase into the gas phase. Since this is also true for natural gas–crude systems, it is possible to describe CO_2–oil phase behavior in traditional reservoir engineering terminology.

CARBON DIOXIDE–OIL SOLUBILITY, R_{so}. The high solubility of CO_2 in crude oil makes it an attractive immiscible flooding prospect. As already mentioned, this affinity for CO_2

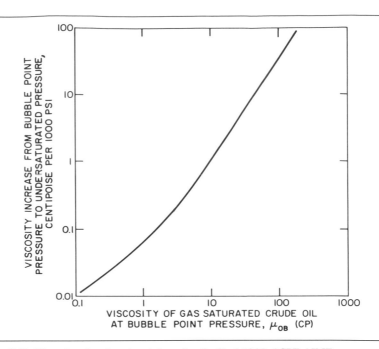

Figure 3.23 Viscosity of undersaturated crude oils (Beal 1946) ©SPE-AIME.

leads to swelling of the oil and viscosity reduction. A solubility correlation has been developed by Simon and Graue (1965) and is presented in figure 3.24A. In this figure, solubility is expressed as X_{CO_2}, the mole fraction of CO_2 in a CO_2-oil mixture. For a given saturation pressure and reservoir temperature, X_{CO_2} can be converted to R_{so} knowing that, by definition,

$$R_{so} = \frac{X_{CO_2}}{(1 - X_{CO_2})} \left(\frac{\text{moles } CO_2}{\text{moles oil}}\right). \tag{3.17}$$

First, convert the moles of CO_2 to SCF by multiplying by 379.4 (assuming a standard T and P of 60°F and 14.7 psia). Then, convert moles of oil to stock tank barrels by, (1) if unknown, estimating oil molecular weight using Cragoe's (1957) estimates:

$$M_{oil} = \frac{6{,}084}{\gamma_o - 5.9}, \tag{3.18}$$

and (2) estimating oil density at standard conditions by definition:

$$\rho_{osc} = (62.4)\left(\frac{141.5}{131.5 + \gamma_o}\right) \tag{3.19}$$

where

M_{oil} = molecular weight of stock tank oil.
γ_o = oil gravity, °API.
ρ_{osc} = stock tank oil density, #/ft³.

Therefore,

$$R_{so} = \frac{(X_{CO_2})(379.4)}{(1 - X_{CO_2})(M_{oil})/[(\rho_{osc})(5.615)]} \qquad (3.20)$$

Figure 3.24C gives the solubility correction, factor (multiply by X_{CO_2} from fig. 3.24A) for oils whose K characterization factors differ from 11.7. These characterization factors (Watson et al. 1935), a measure of crude oil's light and distillable components, can be determined using figure 3.24B, the viscosity, and the API gravity of the oil.

Care should be taken, however, when using the Simon and Graue (1965) correlations. They are empirical fits for use in making preliminary observations. They should not be used as a substitute for true laboratory measurements when making final predictions.

CARBON DIOXIDE–OIL FORMATION VOLUME FACTOR, B_o. The swelling of CO_2-oil systems has been pointed out previously as a major recovery factor in miscible and immiscible CO_2 flooding. Work by Simon and Graue (1965) correlates this effect as a function of pressure, temperature, and oil composition and is shown in figure 3.25.

The mole fraction of CO_2, (X_{CO_2}), can be found as before, using figures 3.24A, B, and C. Second, if the molecular weight of the stock tank oil is unknown, Cragoe's (1957) estimate can be used. Last, the density of the stock tank oil, (ρ_{osc}), is in gm/cc and should be calculated to determine the factor $[M_{oil}/\rho_{osc}]$.

CARBON DIOXIDE–OIL DENSITY ρ_o. The density of CO_2-swollen oil can be represented by

$$\rho_o = \frac{(\rho_{osc})(5.615) + \dfrac{(R_{so})(M)}{(379.4)}}{(B_o)(5.615)} = \frac{\rho_{osc} + \dfrac{(R_{so})(M)}{2130.3}}{B_o}, \qquad (3.21)$$

where

ρ_{osc} = stock tank oil density at standard conditions, #/ft³,
ρ_{osc} = (62.4)(141.5)/(131.5 + °API),
R_{so} = gas solubility, SCF/STB,
M = molecular weight of gas,
B_o = formation volume factor, bbl/STB,
ρ_o = saturated oil density, #/ft³.

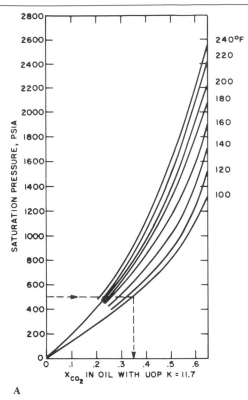

A

Figure 3.24A Solubility of CO₂ in crude oil as determined by pressure and temperature.

Figure 3.24 Generalized correlations for solubility of CO₂-crude oil systems (Simon and Graue 1965) ©SPE-AIME.

CARBON DIOXIDE–OIL VISCOSITY, μ_o. Simon and Graue (1965) have presented a correlation to determine the viscosity of a CO₂-saturated crude. These data are plotted in figure 3.26 for a reservoir temperature of 120°F and show a decrease in oil viscosity of over 90% in many instances. Additional correlations for other temperatures are provided in the Simon and Graue text. If μ_o, the gas-free viscosity at 120°F, is unknown, figure 3.21 can be used to make an estimate of its value.

3.4   INTERMEDIATE PRESSURE, HIGH TEMPERATURE (>122°F) APPLICATIONS. At reservoir pressures higher than those in Region I but lower than Region IV of figure 3.10, supplemental production mechanisms come into play. In addition to increasing reservoir pressure, swelling oil, and viscosity reduction, hydrocarbons may be vaporized into the gas phase.

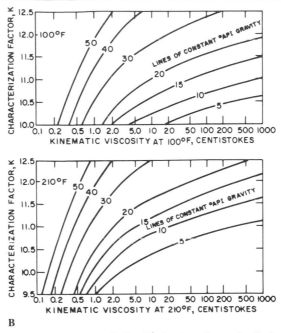

B

Figure 3.24B Characterization factor (K) for C_7^+ fraction of a crude oil given the oil viscosity and gravity.

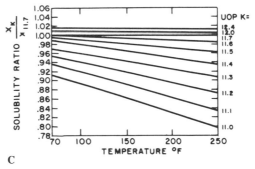

C

Figure 3.24C Correction factor for solubility of CO_2 in crude oil as a function of the characterization factor (K).

For a given crude oil and reservoir temperature, CO_2 extracts oil in increasing amounts with increasing pressure, as clearly shown in the preceding section. This swelling of the crude oil occurs only to a point. After a given pressure, the oil begins to vaporize into a CO_2-rich gas phase. This swelling and subsequent vaporization is shown by Holm and Josendal (1982) in figure 3.27.

At 1,400 psia, the Mead-Strawn crude has swollen to 40% of its original volume. Above 1,400 psia, the oil phase begins to shrink significantly, indicating hydrocarbon

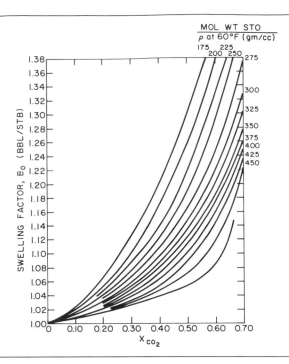

Figure 3.25 Oil formation volume factor for CO_2-saturated crude oils (Simon and Graue 1965) ©SPE-AIME.

extraction. Note that as vaporization occurs, there is a sizeable increase in oil recovery until at flood pressures of 2,400 psia, essentially 100% of the oil in place is recovered.

In similar experiments, Menzie and Nielsen (1963) determined that crude oil could be produced successfully by a process of crude oil vaporization using CO_2. Crude oil was introduced into a windowed cell and charged with CO_2. The vapor was drawn off and the oil was recontacted with CO_2. They found, as shown in figure 3.28, that over half of the original stock tank oil could be vaporized during multiple contacts with CO_2. They also discovered that vaporization of crude oil was a strong function of pressure and that little extraction of reservoir fluid occurred under 1,000-psia saturation pressure.

Data from the Weeks Island pilot test (Perry 1980) also indicate that significant amounts of oil were recovered as a result of vaporization. Note that while this recovery mechanism alone is not usually sufficient to justify project development, overall recoveries are augmented by crude oil vaporization.

Holm and Josendal (1982) report that this extraction of liquid hydrocarbons into a CO_2-rich vapor phase occurs when the density of CO_2 is at least 0.25 to 0.35 gm/cc. The corresponding reservoir temperature-pressure pair that yields such densities for vaporization to occur is shown in figure 3.29. These values of P and T match those presented as the lower limit of Region II initially shown in figure 3.10.

However, the minimum density at which sufficient hydrocarbon extraction takes place to give approximately 94% recovery is about 0.42 gm/cc. At this density, CO_2 has

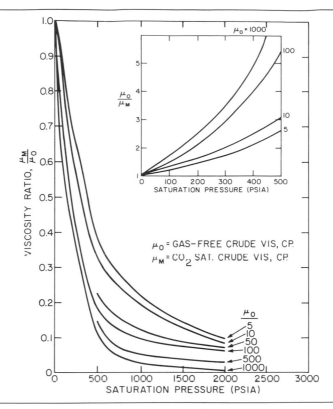

Figure 3.26 Viscosity of carbonated crude oils at 120°F (Simon and Graue 1965) ©SPE-AIME.

enough solvency to dissolve hydrocarbons completely up to C_{30} (Holm and Josendal 1982). Region II-type displacements, then, are bounded by the pressure-temperature regime where CO_2 density is between 0.25 and 0.42 gm/cc (see fig. 3.10). Therefore, when displacement pressure and temperature fall within this region, oil vaporization into the CO_2 vapor phase occurs, but not usually in sufficient quantities to obtain high ultimate recovery.

The injection-vaporization process leads to a gradation of fluids within the cross section of the reservoir ranging from virgin reservoir oil to 100% CO_2 vapor. One useful way of displaying this compositional information is through the use of pressure-composition (*P-X*) diagrams. The diagrams, similar to that presented in figure 3.30, are constructed by injecting various amounts of CO_2 into a windowed cell loaded with reservoir oil and viewing the relative volumes of liquid and vapor present. Note that the diagram represents single contacts of CO_2 with reservoir oil only and presents overall cell compositions rather than individual phase compositions. Even though the *P-X* diagram does not contain all the information of interest, it is a simple and quick method to review the overall phase behavior of a given CO_2/reservoir oil system.

Figure 3.30 shows that the bubble point increases with pressure as CO_2 is added to

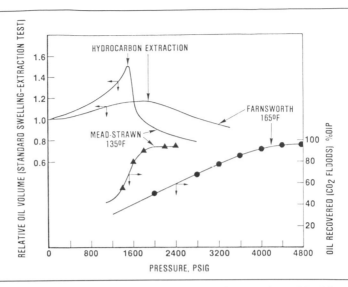

Figure 3.27 Standard swelling/extraction tests and CO₂ flood results on Mead-Strawn and Farnsworth stock tank oils (Holm and Josendal 1982) ©SPE-AIME.

the mixture. It continues to rise until the critical point is reached. Beyond this point is the dew point region. The two-phase region is commonly represented by a series of isoliquid volume lines that radiate from the critical point.

Insight into the CO_2 displacement process can be gained by examining an isobaric path represented by line (1–2) on the pressure-composition diagram (fig. 3.30). Near the beginning of the oil-CO_2 mixing zone, only the original reservoir fluid (represented by 1) passes a given reference point. As the displacement continues, the liquid volume percentage decreases at the reference point in the reservoir until the dew point is reached. After this point, only pure CO_2 (point 2) passes the reference point.

An investigation of this isobaric path is necessary in modeling gas/oil displacement processes. Since there is a transfer of oil from the liquid phase into the vapor, traditional solubility representations (B_o, R_{so}, etc.) of CO_2-oil phase behavior cannot be used to predict fluid properties during the displacement process. An equation of state coupled with a complex reservoir model (see chapter 4) must be used to match the phase behavior as presented in figure 3.30. Such a model is then used to predict fluid properties in the mixing zones between injected and in-place fluids and the efficiency of their displacement.

3.5   INTERMEDIATE PRESSURE, LOW TEMPERATURE (<122°F) APPLICATIONS. At lower temperatures than those discussed in section 3.4, the CO_2 displacement process still exhibits oil swelling, viscosity reduction, increased injectivity, and blowdown recovery. However, reservoirs with pressure-temperature characteristics similar to Region III on figure 3.10 may, rather than vaporize crude oil, extract the crude's lower ends

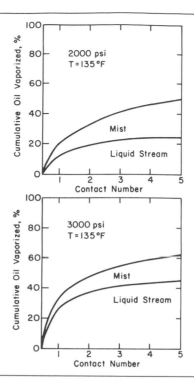

Figure 3.28 Vaporization of stock tank oil by repeat CO₂ contacts (Menzie and Nielsen 1963).

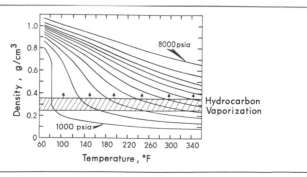

Figure 3.29 Density of CO₂ as a function of pressure and temperature (Holm and Josendal 1982).

forming CO_2-rich liquid mixtures. A number of investigators have observed this phenomenon of multiple liquid phases at low reservoir temperatures ($<122°F$) and pressures greater than 1,000 psia (Gardner et al. 1981; Henry and Metcalfe 1983; Huang and Tracht 1974; Orr and Lien 1980; Rathmell et al. 1971; Shelton and Yarborough 1977).

This development of multiple liquids can be visualized by a re-examination of the

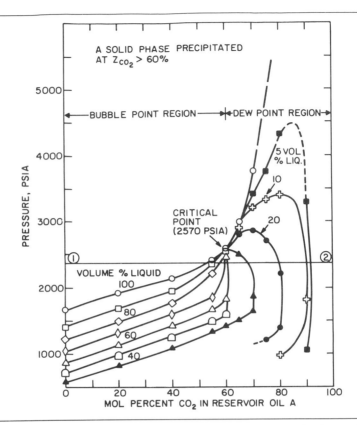

Figure 3.30 Single-contact equilibrium phase behavior for CO$_2$-crude oil mixtures (Simon et al. 1978) ©SPE-AIME.

phase equilibrium diagrams already presented. Figure 3.31 shows the liquid-vapor behavior of a high temperature reservoir on the right. As previously discussed, a cross section of the mixing zone will show one liquid and one vapor present. If the temperature of the system is reduced, the phase diagram on the left occurs.

At high flooding pressures (isobar 1), two liquids will be present at the same point in the mixing zone. With lower pressures (isobar 2), two liquid phases plus one gas phase may be present at the same point in the reservoir. However, multiphase existence does not preclude the ability of CO$_2$ to generate high oil recoveries. In addition to the potential for reduced total mobility in the three-phase region, Orr et al. (1983) report that CO$_2$-rich liquid phases extract more and much heavier hydrocarbons than their rich vapor counterparts.

It is doubtful, however, if the type of data presented in figure 3.31 can be used to predict accurately displacement behavior. Work by Gardner et al. (1981) advocates the addition of multiple-contact experiments to augment single-contact phase equilibrium information. Data from both types of experiments were combined to construct ternary

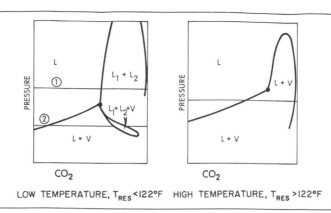

Figure 3.31 Temperature effects on single-contact equilibrium phase behavior.

diagrams (see section 3.6) that were then incorporated in a one-dimensional simulator. They found their simulation results consistent with displacement efficiencies in the laboratory.

3.6   **HIGH PRESSURE MISCIBLE APPLICATIONS.** At high reservoir pressures (>2,000 to 3,000 psia and represented by Region IV of fig. 3.10), CO_2 may vaporize significant quantities of crude oil so rapidly that multiple-contact miscibility occurs in a very brief time period and over a very short reservoir distance. It is normally assumed that CO_2 displacements taking place in Region-IV-type reservoirs obtain instantaneous miscibility; however, because of the high pressures needed, it is doubtful if any field applications of CO_2 will involve true first-contact miscibility.

First-contact miscibility can be represented on the pressure-composition diagram as an isobar that does not pass through a multiphase region. Such a process is shown in figure 3.32 by line 1-2; that is, at the given reservoir pressure and temperature, the reservoir oil (1) and the injected CO_2 (2) are miscible in virtually any proportion.

An alternate way to observe the first-contact miscibility process is through ternary diagrams. A sample diagram is shown in figure 3.33A. This triangular diagram represents mixtures of three pure or pseudocomponents, with each point on the figure representing a given mixture. Also, the data on the ternary diagram represent conditions at a single temperature and pressure, usually a potential reservoir displacement pressure and reservoir temperature.

Suppose component A is added to a mixture of C and D represented by point B, then the resulting composition will lie along a straight line joining A and B. The final composition of a mixture of A and B will depend upon the number of moles of each that is combined. If all mixtures of A and B in any proportion (represented by the dotted line) fall only in the one-phase region, then all mixtures of A and B are miscible on first

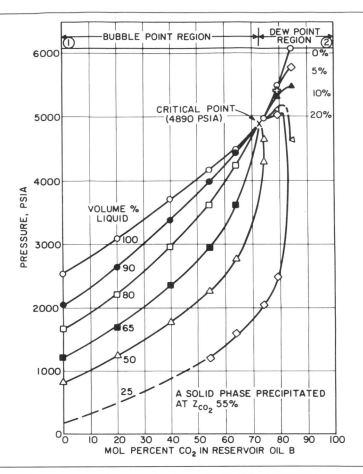

Figure 3.32 Phase diagram of a reservoir oil mixed with CO_2 at 255°F (Simon et al. 1978) ©SPE-AIME.

contact. However, for the case shown, some of the mixtures of A and B fall in the two-phase region, meaning that they will not be miscible but rather separated into two-phases with two distinct compositions. For example, if a mixture of A and B has an overall composition represented by point AB, two phases will be present with compositions at the end of tie lines through point AB. One phase will be an A-rich mixture shown by point a, while the other phase is rich in component D and represented by point d. The amount of each phase present is inversely proportional to the distance from the end points to the overall composition point, AB.

Also note that the tie lines are not usually parallel since the injected gas A does not distribute itself uniformly if two phases are present. For the same reason, the critical, or plait, point is almost never at the peak of the two-phase envelope curve with the slope of the tie lines dependent on the value of the equilibrium coefficient for component A.

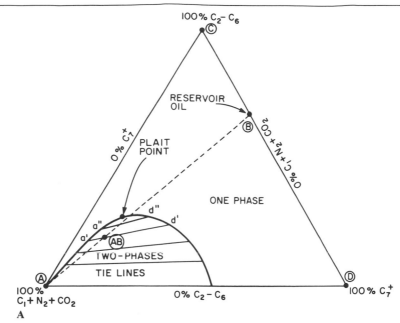

Figure 3.33A Sample ternary phase diagram.

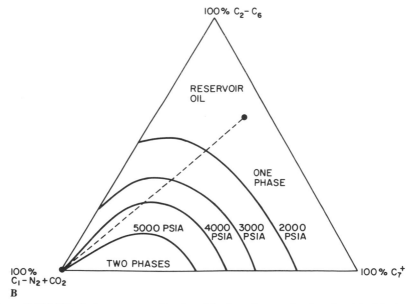

Figure 3.33B Effect of pressure on the size of the two-phase region (envelope) and first-contact miscibility.

As mentioned earlier, first-contact miscibility occurs when a line can be drawn between injected fluid and reservoir oil without passing through the two-phase region. An idealized schematic of this process is shown in figure 3.33B. Note that at a constant reservoir temperature, the size of the two-phase envelope shrinks with increasing displacement pressure. In fact, at very high pressures, the two-phase region may disappear. From this diagram, it is easy to see that a minimum CO_2 displacement pressure of 5,000 psia is needed to achieve first-contact miscibility. At lower flood pressures, oil vaporization or CO_2 condensation may occur, and multiple-contact miscibility may or may not be developed, depending on the relationship between the composition of the reservoir oil, the critical tie line, and the two-phase envelope.

3.6.1   **Multiple-Contact Miscibility.** The determination of whether vaporization or condensation will occur in a CO_2 flood is related directly to the phase equilibriums of the CO_2-reservoir system. At high reservoir pressures and temperatures (Region IV), the mechanism is usually one of crude vaporization. At low temperatures (Region III), the mechanism is best described as the condensation of CO_2 into the oil phase.

When these multiple contacts of CO_2 and reservoir oil develop modified phases that are miscible with each other, the displacement process is defined as multiple-contact miscible. Given the proper phase behavior, these multiple contacts can occur very close to the wellbore and be almost as efficient as true first-contact miscibility.

At high reservoir temperatures, CO_2 extraction of C_2 through C_{30} hydrocarbons into the CO_2 gas phase may after many CO_2-oil contacts, lead to the development of a hydrocarbon-rich gas phase that may be miscible with the uncontacted oil. The process of vaporization of an oil's components is not unlike the high pressure lean gas drive process first introduced by Hutchinson and Braun (1961) and discussed in chapter 1.

The development of miscibility by CO_2 vaporization can be visualized conceptually with a ternary phase diagram constructed from the pseudocomponents, methane-nitrogen-carbon dioxide, ethane through hexane, and heptanes plus. Metcalfe and Yarborough (1979) then give the following discussion of the process. Figure 3.34 presents the two-phase boundary for a particular displacement pressure and temperature, as well as the path predicted by the Hutchinson and Braun theory for a vaporization process. According to Hutchinson and Braun, vapor must be enriched continually by stripping intermediate components from the liquid phase. This enrichment of the vapor continues until a critical composition is reached. The critical composition must be re-established continually by this same process because it is continually destroyed by dispersion effects. Also, note that a critical tie line is shown in figure 3.34. In order for the multiple-contact miscibility vaporization process to be successful, the reservoir fluid composition must lie to the right of the critical tie line. If both fluids lie to the left, vaporization will occur but not in sufficient quantities to develop miscibility. Again, if the injection fluids and original oil both lie to the right of the critical tie line, the fluids are miscible on first contact.

At low reservoir temperatures, CO_2 condensation into a CO_2-rich liquid phase is like enriched gas injection where an oil ring enriched with light hydrocarbons is formed around the wellbore and promotes miscibility between the reservoir oil and the trailing injected gas.

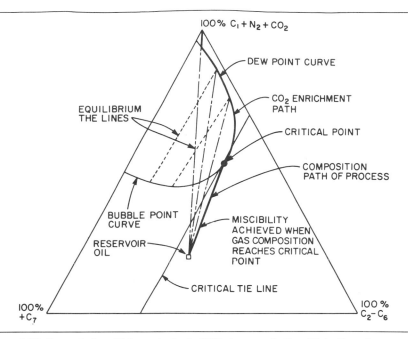

Figure 3.34 Concept of multiple-contact miscibility by vaporization (Metcalfe and Yarborough 1979).

A study by Leach and Yellig (1981) described the condensation process with ternary diagrams. Figure 3.35 represents their multicomponent system. As the displacement progresses, CO_2 enriches the reservoir oil, point C, and the mixture compositions in the mixing zone around the wellbore move along the path from point C to point D. Continued CO_2 injection causes the composition of the wellbore fluid to move from D to B and then from B to A. As the composition path progresses from D to B, the saturated liquid is being enriched with CO_2; hence, the CO_2 is condensing into the oil phase, eventually generating a point of miscibility, point B.

For multiple-contact miscibility by condensation to occur, the injected fluid, A, must be to the right of the critical tie line. If it is not, condensation of CO_2 into the reservoir crude will still occur; however, miscibility will not be developed.

3.6.2 **Correlations to Estimate Miscibility.** Before presenting generalized correlations to determine whether a CO_2 flood will (1) vaporize/condense into the reservoir crude in sufficient quantities to be first-contact or multiple-contact miscible or (2) condense/vaporize and be immiscible, let us first define miscibility.

Individual researches identify a given CO_2 displacement as truly miscible by several criteria, be it the breakpoint in an oil recovery versus pressure curve, 94% total oil recovery, and so forth. Figure 3.36 describes how five research groups have defined minimum miscibility pressure.

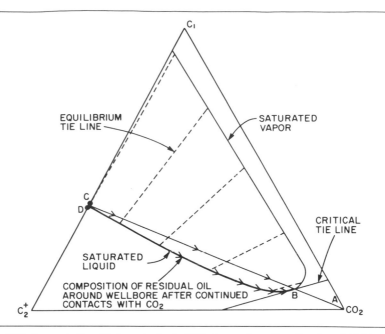

Figure 3.35 Concept of multiple-contact miscibility by condensation (Leach and Yellig 1981).

Regardless of the definition one accepts, a rough formula for estimating the reservoir pressure needed for miscibility, based on oil gravity, reservoir temperature, and reservoir depth, was devised by the National Petroleum Council (1976) and is presented in table 3.1. In determining whether a given field can attain miscibility, use the following estimate to project the maximum reservoir pressure attainable:

$$P_{max} = (0.6D) - 300, \tag{3.22}$$

where

P_{max} = maximum reservoir pressure, psia,
D = reservoir depth, feet.

Two more rigorous studies by Yellig and Metcalfe (1980) and Holm and Josendal (1982) estimate minimum miscibility pressures as functions of reservoir temperature and oil composition. These are shown in figure 3.37. Note that the Yellig and Metcalfe correlation does not account for changes in oil composition while Holm and Josendal's criteria determine miscibility as a function of reservoir temperature and the molecular weight of the C_5^+ oil fraction. Also, Holm and Josendal point out that heavier gravity oil (as C_5^+ increases) demands notably higher minimum miscibility pressures.

A more recent method of determining CO_2 minimum dynamic miscibility pressures (MDMP) has been developed by Johnson and Pollin (1981). This correlation accounts

Yellig and Metcalfe (Amoco) (1980, p. 163)

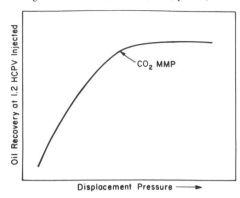

Holm and Josendal (Union) (1982, p. 89)
"80% of the in place oil is recovered at CO_2 breakthrough and 94% at a GOR of 40,000 SCF/bbl."

Williams et al. (Chevron) (1980, p. 67)
"90% oil recovery at 1.2 hydrocarbon pore volumes of CO_2 injected."

Perry (Shell) (1980)
"smooth transition from zero to full light transmittance over a production interval of several percent of a pore volume" in a 5-ft long vertical sand pack run below the critical velocity as defined by Dumoré.

Johnson and Pollin (Phillips) (1981, p. 271)

"breakpoint in the [oil] recovery [versus pressure] curve is clearly identifiable . . . a slim tube miscibility can be defined there."

Figure 3.36 Definitions of minimum miscibility pressure.

for changes in miscibility due to oil gravity, molecular weight, reservoir temperature, and injection gas composition and shows better agreement to experimental results than those methods previously discussed. For reservoir temperatures in the range of 80°F to 280°F, the correlating equation has the form

$$P_{\text{mdmp}} - P_{c,\text{inj}} = \alpha_{\text{inj}}(T_{\text{res}} - T_{c,\text{inj}}) + I(\beta M - M_{\text{inj}})^2, \qquad (3.23)$$

where

P_{mdmp} = predicted minimum dynamic miscibility pressure, psia,
$P_{c,\text{inj}}$ = injection gas critical pressure, psia,
M = average molecular weight of the oil,
β = constant, 0.285,
T_{res} = reservoir temperature, °K,
$T_{c,\text{inj}}$ = injection gas critical temperature, °K,
M_{inj} = molecular weight of injection gas,
α_{inj} = 18.9 psia/°K (for pure CO_2).

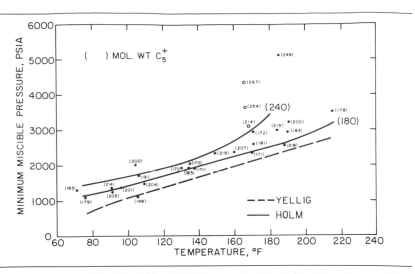

Figure 3.37 Holm and Josendal (1982) and Yellig and Metcalfe (1980) minimum miscibility pressure correlations for carbon dioxide ©SPE-AIME.

For nitrogen-diluted injection gas ($Y_{CO_2} > .9$):

$$\alpha_{inj} = 10.5\left(1.8 + \frac{10^3 Y_2}{T_{res} - T_{c,inj}}\right). \tag{3.24a}$$

For injection gas containing methane ($Y_{CO_2} > .9$):

$$\alpha_{inj} = 10.5\left(1.8 + \frac{10^2 Y_2}{T_{res} - T_{c,inj}}\right), \tag{3.24b}$$

where Y_2 is the mole fraction of the diluting component. Last, I is defined as the crude oil characterization index. An approximation of the index, I, can be obtained from the Watson et al. (1935) K-factor discussed earlier in section 3.3.2 by

$$I = 2.22K - 25.84 + 0.66K^{-2}. \tag{3.24c}$$

This method of determining minimum miscibility pressures for the CO_2 injection process takes into account the findings of Holm and Josendal (1974) that states that higher molecular weight oils have higher miscibility pressures. Johnson and Pollin (1981) have also included experimental findings that other components including nitrogen and methane in the injected CO_2 can alter the pressure needed for miscibility significantly upward by increasing the size of the two-phase region on the ternary diagram.

One example by Stalkup (1983) shows miscibility pressure increasing from 1,200 to 2,000 psia with a 20 mole percent contamination of the injected CO_2 gas with methane. A similar contamination concentration with nitrogen raised the miscibility pressure to 4,200 psia. Hydrogen sulfide in the injected CO_2 stream shows the opposite effect and

Table 3.1 Estimate of CO$_2$-Crude Miscibility Pressure

Miscibility Pressure Versus Oil Gravity

Gravity (°API)	Minimum Miscibility Pressure (psia)
<27	4,000
27–30	3,000
>30	1,200

Correction for Reservoir Temperature

Temperature (°F)	Additional Pressure Required (psia)
120	None
120–150	+ 200
150–200	+ 350
200–250	+ 500

Source: National Petroleum Council, *Enhanced Oil Recovery*, Washington, D.C. (1976).

slightly lowers miscibility pressure. Metcalfe (1982) examined the effect of H$_2$S dilution on miscibility pressure and reported a decrease from 1,200 psia to 1,120 psia as H$_2$S concentration was increased from 0 to 25 mole percent.

A fifth correlation by Cronquist (1978) characterizes the miscibility pressure as a function of reservoir temperature, molecular weight of the reservoir crude's pentanes-plus fraction, and the mole percentage of methane and nitrogen. The relationship takes the form:

$$P_{mdmp} = 15.988 \ (T_{res})^{0.744206 \ + \ 0.0011038(MWC_5^+) \ + \ 0.0015279(Y_{C_1})} \tag{3.25}$$

where

P_{mdmp} = predicted minimum dynamic miscibility pressure, psia,
T_{res} = reservoir temperature, °F,
MWC_5^+ = molecular weight of pentanes-and-heavier fraction,
Y_{C_1} = mole percentage of methane and nitrogen.

Cronquist's correlation is an empirical fit of 58 data points. The oils tested ranged from 23.7 to 44° API with reservoir temperatures ranging from 71 to 248°F. A range of miscibility pressures from 1,075 to 5,000 psi was predicted.

A sixth correlation for impure CO$_2$ streams and live oil systems has been prepared by Alston et al. (1983). Their empirically derived correlation for estimating the minimum pressure required for miscibility uses, as primary correlating factors, reservoir temperature, the reservoir oils pentanes plus molecular weight, the ratio of volatile to intermediate components, and CO$_2$ purity. The correlation takes the form

$$P_{mdmp} = 8.78 \cdot 10^{-4} \ (T)^{1.06}(C_5^+)^{1.78}(VOL/INT)^{0.136}(87.8/T_{cm})^{(170/T_{cm})}, \tag{3.26}$$

where

P_{mdmp} = predicted minimum dynamic miscibility pressure, psia,
T = reservoir temperature, °F,
C_5^+ = pentanes and heavier molecular weight,
VOL/INT = ratio of volatile (C_1 and N_2) to intermediate (C_2–C_4, CO_2 and H_2S) mole
              fractions in the reservoir oil,
T_{cm} = pseudocritical temperature of pure or impure CO_2 injection stream, °F

Note that if the calculated MDMP falls below the reservoir bubble point, the MDMP should be taken as the bubble point pressure.

This pseudocritical temperature of the injection stream can be estimated using a weight fraction mixing rule:

$$T_{cm} = \sum_{i=1}^{n} w_i T_{ci} - 459.7. \tag{3.27}$$

The critical temperatures of the injection stream components used in this calculation for CO_2, N_2, C_1, C_3, and C_4 were the true critical temperatures (°R) for these components. However, a critical temperature of 585°R was used for H_2S and C_2 impurities in the CO_2 stream.

Alston et al.'s (1983) work compares the values of 68 experimentally determined MDMPs. Thirty-two of the predicted MDMP values were higher than the experimental values, while 35 were lower and an average error of 177 psia was encountered. The maximum and minimum predicted MDMPs were 5,166 and 985 psia respectively, with a maximum error of 803 psia.

Stalkup (1983) has compared the predicted values of four of the correlations against 19 experimentally determined miscibility pressures. His work is shown in figure 3.38. It should be noted that the Yellig and Metcalfe (1980) and Johnson and Pollin (1981) correlations usually predicted too low a miscibility pressure while the Cronquist (1978) method was high for 16 of the 19 oils. The Holm and Josendal (1982) and Alston et al. methods predicted an even split of high and low answers.

Although these correlations are helpful as a screening tool, they are not adequate for final design. The best way of determining the pressure needed for miscibility is by laboratory tests in which CO_2 displaces reservoir crude from slim tube sand packs. A more thorough treatment of these laboratory displacement procedures occurs later in this section.

3.7   GAS-WATER INTERACTION PROPERTIES

3.7.1   A Review of Water–Natural Gas Systems

SOLUBILITY OF NATURAL GAS IN CONNATE WATER. The solubility of methane in interstitial water as a function of temperature and pressure was investigated by Culber-

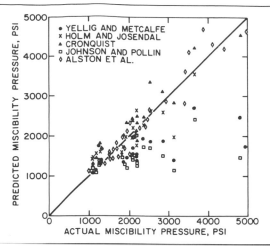

Figure 3.38 Accuracy of screening guides for predicting CO_2 miscibility pressure (Stalkup 1983) ©SPE-AIME.

Correlation	Average Error (PSIA)	Maximum Error (PSIA)
Yellig and Metcalfe (1980)	580	2,370
Holm and Josendal (1982)	280	1,160
Cronquist (1978)	330	850
Johnson and Pollin (1981)	790	3,400
Alston, et al.[a] (1983)	177	803

[a]Not same data as other correlations.

son and McKetta (1951). Their experimental results are shown in figure 3.39 (A) and (B). The first plot represents the solubility of the gas in pure water, while the second plot gives the correction factors necessary to account for the decrease in gas solubility with increasing salinity of the water.

These results permit certain generalizations concerning the solubility of natural gas in water and brine:

1. The solubility of natural gas in connate water is small compared to its solubility in crude oil at comparable temperatures and pressures. An examination of figure 3.39 shows that the solubility at 2,000 psia is on the order of 12 SCF per barrel. At this pressure the solubility of natural gas in crude oil is about 900 SCF per barrel for a 50° API crude.

2. At constant temperature the solubility of natural gas in connate water increases with pressure. However, the solubility is not a linear function of pressure as required by Henry's Law.

3. At constant pressure the solubility initially decreases with temperature. However, at high pressures the solubility reaches a minimum so that a further increase in temperature brings about an increase in solubility. As already pointed out, this effect is not

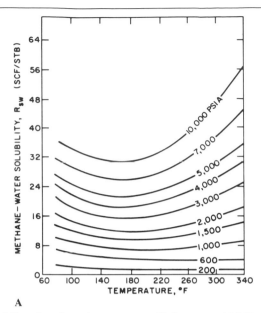

A

Figure 3.39A Solubility of methane in pure water (Culberson and McKetta 1951).

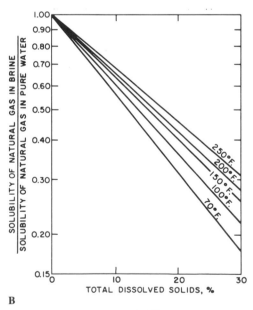

B

Figure 3.39B Effects of dissolved salts on the solubility of natural gas in water (Culberson and McKetta 1951).

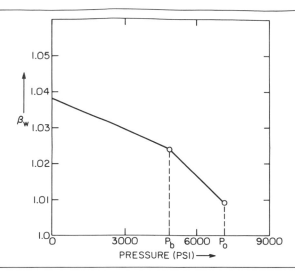

Figure 3.40 Typical plot of the water formation volume factor as a function of pressure at reservoir temperature.

observed in natural gas–crude oil systems in the temperature and pressure ranges investigated. However, these minima in solubility are not uncommon for gases that are only slightly soluble in a liquid.

4. The solubility of natural gas in brine decreases with increasing salinity of the brine.

WATER-FORMATION VOLUME FACTOR. A typical plot of B_w as a function of pressure is shown in figure 3.40. If the water is initially undersaturated, as the pressure is decreased from P_o to P_b, then the value of B_w increases due to the expansion of the liquid. At pressures below the bubble point (P_b), gas is evolved, but because of the low solubility of natural gas in brine, the shrinkage of the liquid phase is relatively small. This shrinkage is usually insufficient to counterbalance the expansion of the liquid on release of pressure so that B_w continues to increase below the saturation pressure as shown. However, the rate of increase with decreasing pressure is lower at pressures below P_b than at pressures above P_b. It is also noteworthy that in the pressure ranges most commonly encountered in petroleum reservoirs, the value of B_w does not differ greatly from one. This also is a consequence of the low solubility of natural gas in brine.

Experimental values of the water-formation volume factor below the saturation pressure are shown in figure 3.41. This chart was prepared by Burcik (1956) from data obtained by Dodson and Standing. The curves shown in figure 3.41 represent B_w as a function of pressure at constant temperature for pure water saturated with natural gas. The decrease in B_w with increasing pressure is clearly shown in this diagram, as is the fact that B_w is only slightly different from one in the pressure range investigated.

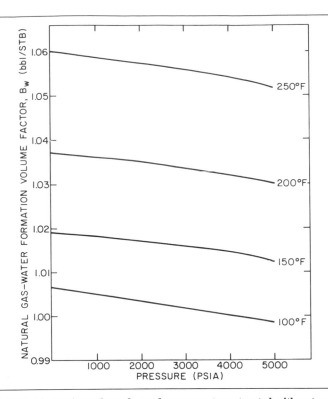

Figure 3.41 Water formation volume factor for pure water saturated with natural gas at four different temperatures (Burcik 1956).

WATER DENSITY. The density of natural-gas-saturated water can be defined as

$$\rho_w = \frac{(\rho_{w_{sc}})(5.615) + \dfrac{(R_{sw})(M)}{(379.4)}}{(B_w)(5.615)} = \frac{\rho_{w_{sc}} + \dfrac{(R_{sw})(M)}{2130.3}}{B_w}, \tag{3.28}$$

where

B_w = formation volume factor, bbl/STB,
$\rho_{w_{sc}}$ = unsaturated water density at standard P and T, #/ft³,
R_{sw} = gas solubility, SCF/STB,
M = molecular weight of gas,
ρ = saturated water density, #/ft³.

Data for B_w and R_{sw} have been presented in previous sections. The value of $\rho_{w_{sc}}$ can be determined using figure 3.42. Calculations of R_{sw} and B_w as a function of pressure then can be performed to build a table of water density, ρ_w.

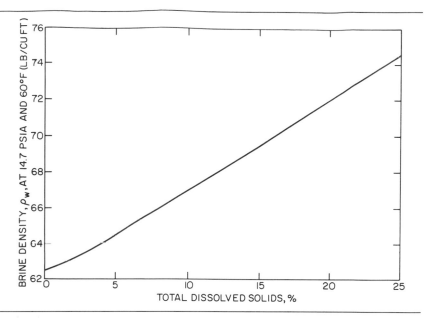

Figure 3.42 The effect of total dissolved solids on the density of brine at standard temperature and pressure.

WATER COMPRESSIBILITY. Values of the water-formation volume factor and density at pressures above the saturation pressure are determined by the coefficient of compressibility of the gas-saturated connate water. Values of these coefficients are given in figure 3.43. Figure 3.43A gives the coefficient of compression of pure water or a gas-free brine as a function of temperature and pressure. Figure 3.43B gives the multiplicative correction factor that must be applied to account for the increase in compressibility due to solution gas (Burcik 1956).

It should be apparent that the coefficient of compression of a given gas-saturated water may be estimated at any temperature and pressure with the aid of the data in figure 3.43. Consequently, the average coefficient may be estimated for any pressure range above the saturation pressure. Values of B_w and ρ_w above the saturation pressure may then be computed using the equations

$$B_w = B_{wb} \exp[c_w(P_b - P)] \tag{3.29a}$$

and

$$\rho_w = \rho_{wb} \exp[c_w(P - P_b)], \tag{3.29b}$$

where B_w and ρ_w are the values of the water-formation volume factor and density above the bubble point pressure at pressure P, subscript b denotes the value of the water-formation volume factor and density at the saturation (bubble point) pressure P_b, and c_w is the average coefficient of compression over the pressure range P to P_b.

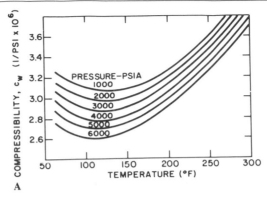

Figure 3.43A Compressibility of water as a function of pressure and temperature (Burcik 1956).

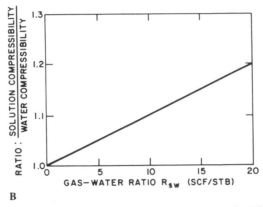

Figure 3.43B Effect of dissolved gas on water compressibility (Burcik 1956).

VISCOSITY OF CONNATE WATER. The effect of high pressures on the viscosity of pure water was investigated by Bridgeman (1931). Some of his results are reproduced in figure 3.44. From an examination of the data presented, it is immediately apparent that the change in water viscosity with pressure is small for the pressure range usually encountered in petroleum engineering practice.

The addition of salts to water causes an increase in viscosity, as shown in figure 3.45. The data presented are for sodium chloride solutions containing 0, 5, 10, 15, 20, and 25 gm salt per 100 gm water. This figure also shows the dependence of viscosity on temperatures for these solutions (Burcik 1956).

It has been pointed out that small amounts of natural gas dissolve in connate water at reservoir temperature and pressure. However, no data on the effect of this solution gas on water viscosity have been published. Undoubtedly, the solution gas causes a decrease in the water viscosity, but the magnitude of this decrease is unknown.

In view of the lack of experimental data, it is impossible to make an accurate estimate

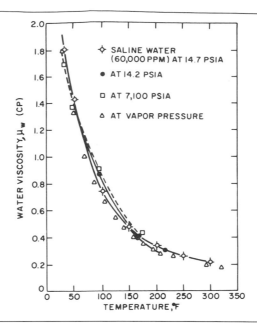

Figure 3.44 Viscosity of water at oil field temperature and pressure (Bridgeman 1931).

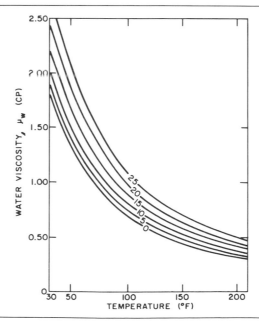

Figure 3.45 Brine viscosity as a function of temperature and solutions containing 0, 5, 10, 15, 20, and 25 gm salt per 100 gm water (Burcik 1956).

of connate water viscosity under reservoir conditions. Consequently, if an experimental value for the brine viscosity is not known, it is common practice to assume a value equal to that of pure water at atmospheric pressure and at reservoir temperature (see curve for no sodium chloride in fig. 3.45). This tacitly assumes that the viscosity of brine is independent of pressure and that the increase in viscosity caused by dissolved salts is more or less counterbalanced by the probable decrease caused by solution gas.

3.7.2   Carbon Dioxide–Water Systems

CARBON DIOXIDE–WATER SOLUBILITY, R_{sw}. The solubility of CO_2 in resident water is an important factor that cannot be neglected in the simulation process. This is especially true when CO_2 is being injected into a previously waterflooded reservoir or when CO_2 is being injected with water (WAG). A thorough study of CO_2 dissolved in distilled water was presented by Dodds et al. (1956) and is shown in figure 3.46. An increase in salinity of the reservoir water decreases the gas solubility significantly. A correction factor combining the works of Crawford et al. (1963), Holm (1963), Johnson et al. (1952), and Martin (1951) is shown in figure 3.47.

CARBON DIOXIDE–WATER FORMATION VOLUME FACTOR, B_w. Swelling data are not available for CO_2-water systems. However, B_w can be derived using the following relationship:

$$B_w = \frac{(\rho_{w_{sc}})(5.615) + \dfrac{(R_{sw})(M)}{(379.4)}}{(\rho_w)(5.615)} = \frac{\rho_{w_{sc}} + \dfrac{(R_{sw})(M)}{2130.3}}{\rho_w} \tag{3.30}$$

where

B_w = formation volume factor, bbl/STB,
$\rho_{w_{sc}}$ = unsaturated water density at standard P and T, #/ft^3,
R_{sw} = gas solubility, SCF/STB,
M = molecular weight of gas,
ρ_w = saturated water density, #/ft^3.

Data for the density of CO_2-saturated water, ρ_w, are presented in a later section.

CARBON DIOXIDE–WATER DENSITY, ρ_w. Extensive water density measurements of CO_2-saturated solutions have been presented by Parkinson and de Nevers (1969). Note that this work, shown in figure 3.48 has a maximum pressure of only 500 psia. However, linear extrapolation of these data to 65 atm shows close agreement ($<1\%$) with work by Francis (1959).

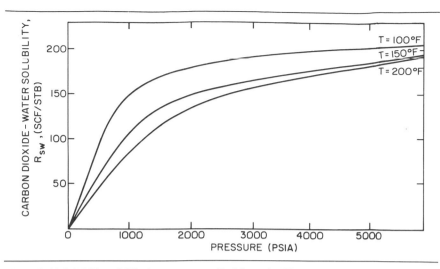

Figure 3.46 Solubility of CO$_2$ in pure water (Dodds et al. 1956).

WATER VISCOSITY, μ_w. Liquid viscosities are usually a strong function of pressure and temperature. However, as previously illustrated, water presents an anomaly, increasing less than twofold while undergoing a pressure increase from 14.7 to 7,100 psi. As a function of temperature, though, water viscosity is found to be strongly dependent. Hawkins et al. (1940) presented the relationship as follows:

$$\mu_w = \frac{2.185}{0.04012T + 0.0000051547T^2 - 1}, \tag{3.31}$$

where

μ_w = water viscosity, cp,
T = water temperature, °F.

The effect of carbonation on water viscosity was investigated by Tumasyan et al. (1969). Their report showed that CO$_2$ solutions prepared in filtered tap water at 1,422 and 2,130 psia and 20°C increased the water viscosity 18.9% and 27.3% respectively. This effect is shown in figure 3.49. There is little information, however, to infer that these data can be extrapolated to other temperatures.

3.8   ROCK-FLUID PROPERTIES

3.8.1   **Rock Compressibility, c_r.** When fluid is removed from a reservoir, the internal pressure is reduced while the external (overburden) pressure remains constant. This being the

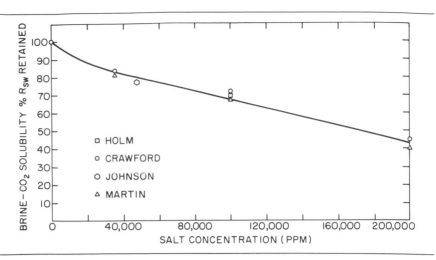

Figure 3.47 Effect of dissolved salts on the solubility of CO_2 in reservoir brines (Holm 1963; Crawford et al. 1963; Johnson et al. 1952; Martin 1951).

case, the reservoir is compacted and porosity decreases. Also, as sand pressure decreases, the individual sand grains expand, causing a further decrease in porosity. This change in reservoir porosity can be represented by the equation

$$\phi = \phi_{in}[1 - c_r(P_{in} - P)], \qquad (3.32)$$

where

P = reservoir pressure, psia,
P_{in} = initial reservoir pressure, psia,
ϕ_{in} = initial porosity, fraction,
ϕ = porosity at P, fraction,
c_r = rock compressibility, PV/PV/psi.

Early work by Hall (1953) and later refinements by Fatt (1958) provided measurements of rock compressibility as a function of initial porosity and net overburden pressure. Hall's work is presented in figure 3.50.

Even though these rock compressibilities are small, on the order of $5 \cdot 10^{-6}$ psi^{-1}, it is important to retain porosity variation. This is especially true above the bubble point where water and oil compressibilities approach rock values and gas saturation is zero. Craft and Hawkins (1959) have shown an error in reserve calculations of over 60% when rock and fluid compressibilities are neglected above the bubble point.

3.8.2   **Relative Permeabilities, k_{ro}, k_{rw}, k_{rg}.** In rocks that are saturated by more than one fluid, competition for flow channels exists. The true permeability that each fluid sees is its

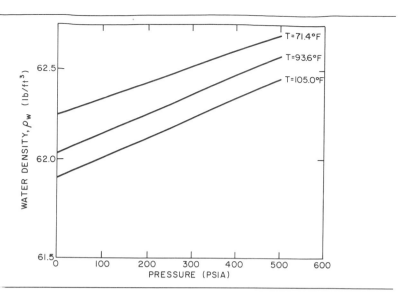

Figure 3.48 Density of equilibrium CO$_2$-water mixtures at three different temperatures (Parkinson and de Nevers 1969).

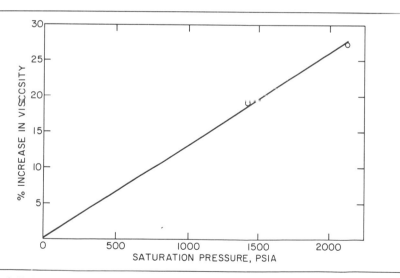

Figure 3.49 Effect of dissolved CO$_2$ on water viscosity (Tumasyan et al. 1969).

relative permeability multiplied by the absolute permeability. Because of this interference between the saturating fluids, the sum of the relative permeabilities is always less than one.

Two-phase relative permeability data are usually generated in the laboratory by simultaneously flowing fluids through a core and measuring saturations, pressures, and

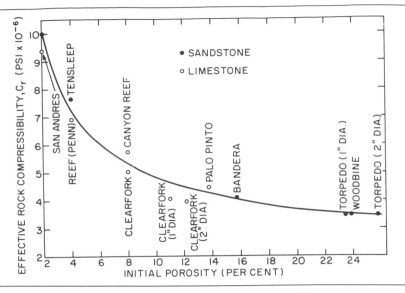

Figure 3.50 Rock compressibility as a function of porosity (Hall 1953).

rates of flow. However, if three-phase data are not available, they can be generated using Stone's (1973) correlation. Stone has developed a model in which three-phase relative permeabilities are estimated from two-phase data. The required data, shown in figure 3.51, consist of a set of oil-water relative permeability and oil-gas relative permeability curves. Water, k_{rw}, and gas, k_{rg}, relative permeabilities are read directly from figure 3.51, while the oil relative permeability, k_{ro}, is calculated explicitly from:

$$k_{ro} = (k_{row} + k_{rw})(k_{rog} + k_{rg}) - (k_{rw} + k_{rg}). \quad (k_{ro} \geq 0) \quad (3.33a)$$

One modified form of Stone's model has been suggested by Aziz and Settari (1979):

$$k_{ro} = k_{rocw}[(k_{row}/k_{rocw} + k_{rw})(k_{rog}/k_{rocw} + k_{rg}) - (k_{rw} + k_{rg})] \quad (3.33b)$$

Here, as before, k_{row}, k_{rog}, k_{rw}, and k_{rg} are read directly from the water-oil and gas-oil relative permeability, respectively, while k_{rocw} is the relative permeability to oil at connate water saturation. Note that a number of additional modifications to Stone's original work are available, including hysteresis effects that should be accounted for in gas-water displacements.

3.8.3   **Interfacial Tension Effects.** Under immiscible conditions, relative permeability curves usually exhibit considerable curvature as shown in figure 3.51, and residual saturations of the respective phases are present. As interfacial tension approaches zero, residual phase saturations decrease toward zero, and the relative permeability curves approach

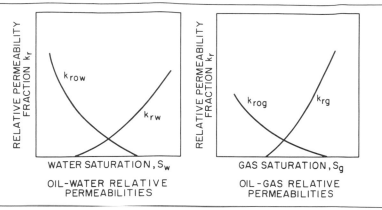

Figure 3.51 Sample relative permeability curves for Stone's (1973) Model.

straight lines. Bardon and Longeron (1980) have demonstrated these effects in the laboratory, and their results are shown in figure 3.52.

While there is no mathematically documented correlation between interfacial tension and gas-oil relative permeability, Coats (1980) and Ngiem et al. (1981) have presented mathematical treatments not based on any theory but simply a device to approximate the described behavior between interfacial tension and relative permeability. Coats's work follows. The *two-phase gas and oil curves* used in his work are

$$k_{rg} = k_{rgcw} \{f(\sigma)\overline{S}_g{}^{n_g} + [1 - f(\sigma)]\,\overline{S}_g\} \tag{3.34a}$$

$$k_{rog} = f(\sigma)\overline{S}_o{}^{n_{og}} + [1 - f(\sigma)]\,\overline{S}_o, \tag{3.34b}$$

where

$$\overline{S}_g = \frac{S_g - S_{gr}^*}{1 - S_{wir} - S_{gr}^*}, \tag{3.34c}$$

$$\overline{S}_o = \frac{1 - S_g - S_{wir} - S_{org}^*}{1 - S_{wir} - S_{org}^*}, \tag{3.34d}$$

and

$$f(\sigma) = \left(\frac{\sigma}{\sigma_o}\right)^{1/n_1} \tag{3.34e}$$

By definition:

σ = interfacial tension,
σ_o = initial interfacial tension,
n_1 = read in exponent (ranges from 4 to 10),

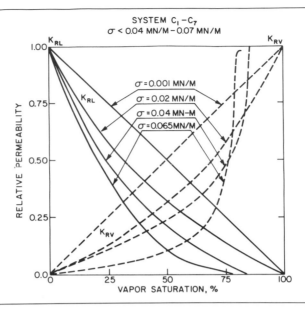

Figure 3.52 Vapor/liquid relative permeabilities for low interfacial tension values (Bardon and Longeron 1980) ©SPE-AIME.

S_{gr}	= residual gas saturation at original gas/oil interfacial tension,
S_{org}	= residual oil saturation to gas at original gas/oil interfacial tension,
S_{wir}	= irreducible water saturation,
S_{orw}	= residual oil saturation to water,
k_{rgcw}	= relative permeability to gas at connate water saturation,
k_{rwro}	= relative permeability to water at residual oil saturation,
n_w, n_{ow}, n_{og}, n_g	= exponents on relative permeability curves.

As interfacial tension decreases, S^*_{gr} and S^*_{org} approach zero as

$$S^*_{gr} = f(\sigma)S_{gr} \tag{3.34f}$$

$$S^*_{org} = f(\sigma)S_{org}. \tag{3.34g}$$

For large n_1 values as σ decreases below σ_o, the value of $f(\sigma)$ will remain near 1 until σ/σ_o is very small. This means that k_{rg} and k_{rog} will vary little with interfacial tension until close proximity to the critical point is attained. This reflects the information of the literature on low tension behavior, which indicates that very low interfacial tensions are necessary to reduce residual oil saturations appreciably under normal reservoir pressure gradients.

For Coats's (1980) *two-phase, oil-water relative permeabilities*, the following analytical relationships were used:

$$k_{rw} = k_{rwro}\left(\frac{S_w - S_{wir}}{1 - S_{wir} - S_{orw}}\right)^{n_w} \tag{3.34h}$$

and

$$k_{row} = \left(\frac{1 - S_w - S_{orw}}{1 - S_{wir} - S_{orw}}\right)^{n_{ow}} \tag{3.34i}$$

Hence, k_{rg} and k_{rw} are calculated directly from equations 3.34(a) and 3.34(h), while the oil relative permeability is estimated using Stone's (1973) modified method presented in the previous section.

3.8.4   **Asphaltene Precipitation.** When CO_2 and reservoir oil mix, it was shown earlier that multiple liquid phases may exist in equilibrium. In some instances, a fourth, solid phase may appear. These heavy ends of a crude oil become less soluble with the addition of CO_2 and precipitate, causing loss of permeability. The amount of precipitate and its subsequent effect on recovery is difficult to quantify.

In addition, CO_2-water mixtures are slightly acidic (carbonic acid) and react accordingly with carbonate portions of the reservoir rock. Besides dissolving the formation matrix, thereby increasing permeability, it is also a well-known fact that some acids, when in contact with crude oil, may precipitate asphaltenes (David 1972).

Crude oil normally consists of a conglomeration of waxes, resins, asphaltenes, and other semisolid materials as listed in table 3.2. Depending on the crude oil characteristics, this can be a major design problem. One should thoroughly examine this effect in the laboratory prior to field implementation.

3.9   **MODELING CARBON DIOXIDE–OIL PHASE BEHAVIOR.** In order to model a CO_2 injection process accurately, some method is required to represent analytically the phase equilibrium that exists between CO_2 and crude oil. For low pressure, Region I, or very high pressure, first contact miscible displacements, little more than traditional PVT representations (B_o, μ_o, R_{so}, etc.) may be required. A more applicable simulation model requires a phase behavior model that is capable of predicting CO_2-crude oil equilibriums over the entire range of possible CO_2 applications. Although convergence pressure models have been used to predict CO_2 crude oil equilibrium (Nolen 1973; Perry 1980; Spivak et al. 1975), they suffer from several disadvantages and limitations. Since a great deal of parameter adjustment is required to match CO_2-crude oil equilibriums, and since this adjustment is system specific, convergence pressure models are not acceptable predictive tools for CO_2 systems. In addition, they are not applicable to liquid-liquid or three-phase, liquid-liquid-vapor, equilibriums, making them unsuitable for low temperature systems.

Table 3.2 Published Data for Crude Oils Which Exhibit Asphaltene Precipitation

Location	Country	Percentage by Weight Asphaltene
Abo Formation, New Mexico	United States	0.1
Balachany	USSR	0.5
Baxterville, Mississippi	United States	17.2
Beaver Creek Field, Madison Formation, Wyoming	United States	1.5
Belridge, California	United States	5.1
Goose River, Beaverhill Lake Formation, Alberta	Canada	1.3
Grozny 1	USSR	0.9
Grozny 2	USSR	1.5
Hould, Texas	United States	0.5
Huntington Beach, California	United States	4
Kalvga	USSR	0.5
Kirkuka	Iraq	1.3
Bibi-Eibat	USSR	0.3
Mexia, Texas	United States	1.3
Mississippi Formation, Oklahoma	United States	2
Oklahoma City, Oklahoma	United States	0.1
Panuco, Mexico	Mexico	12.5
Poza, Mexico	Mexico	2.4
Prudhoe Bay, Alaska	United States	1.4
Redfork Formation, Oklahoma	United States	0.1
Rozet Field, Minnelusa Formation, Wyoming	United States	0.1
Santos and Broudna Formation, California	United States	3.8
Snipe Lake Pool, Beaverhill Formation, Alberta	Canada	2.6
Swan Hill Formation, Alberta	Canada	2.8

Source: A. David, "Asphaltene Flocculation During Solvent Stimulation of Heavy Oils," American Institute of Chemical Engineers, Dallas (1972).

Recent works (Baker et al. 1981; Fussell 1979; Mehra et al. 1977; Turek et al. 1980) have introduced generalized equations of state (EOSs) to model and predict CO_2–crude oil phase behavior such as that presented on P-X and triangular diagrams, and have shown that the EOS approach is not only capable of predicting and modeling the high temperature vapor-liquid-type equilibrium but also can handle the low temperature, three-phase systems.

An EOS is an algebraic relationship between pressure, temperature, and molar volume for a single component or a mixture. The simplest (and most commonly recognized) EOS introduced earlier in this chapter is the imperfect gas law:

$$Z = \frac{PV}{RT} \tag{3.35}$$

As pointed out by Oellrich et al. (1981), the two-parameter-type EOSs are based on the Van der Waals Equation:

$$Z = \frac{V}{V - b} - \frac{a}{RTV},$$ (3.36)

where the parameters a and b are supposed to account for the interaction forces between molecules and for the actual volume of the molecules. For the Van der Waals EOS,

$$a = \frac{26}{64} \frac{R^2 T_c^2}{P_c}$$ (3.37)

and

$$b = \frac{RT_c}{8P_c}.$$

A major improvement in the accuracy of the Van der Waals Equation was achieved by the Redlich-Kwong (1949) modification:

$$Z = \frac{V}{V - b} - \frac{a}{RT^{1.5}(V + b)},$$ (3.38)

with

$$a = \frac{\Omega_a R^2 T_c^{2.5}}{P_c},$$ (3.39a)

$$b = \frac{\Omega_b RT_c}{P_c},$$ (3.39b)

$$\Omega_a = 0.42748,$$ (3.39c)

$$\Omega_b = 0.08664.$$ (3.39d)

In order to calculate mixture properties, the pure component parameters a and b in equation (3.38) are replaced with mixture values given by

$$a_M = \Sigma_i \Sigma_j (x_i x_j a_{ij})$$ (3.40)

and

$$b_M = \Sigma_i (x_i b_i).$$

The a_{ij}s in (3.40) are given by

$$a_{ij} = \sqrt{a_{ii} a_{jj}} (1 - k_{ij}),$$ (3.41)

where the binary interaction coefficients k_{ij} are developed from experimental data on two-component systems.

Significant improvements in the Redlich-Kwong EOS have been made by making the

Ω_a and Ω_b parameters (which are constants in the original Redlich-Kwong EOS) functions of temperature and of one or more characteristic parameters of the individual components. Zudkevitch and Jaffe (1970) proposed calculating Ω_a and Ω_b for each component at a given temperature using the component's saturation pressure, saturated liquid density, and Lyckman's fugacity coefficient. The Soave (1972) modification to the Redlich-Kwong EOS has the form

$$Z = \frac{V}{V - b} - \frac{a}{RT(V + b)},\tag{3.42}$$

with b given by the equations (3.39h) and (3.39d) and

$$a = 0.42747 \frac{R^2 T_c^2}{P_c} \alpha(T_r, \omega),\tag{3.43a}$$

$$\alpha(T_r, \omega) = [1 + m(1 - \sqrt{T_r})]^2,\tag{3.43b}$$

$$m = 0.48 + 1.574\omega - 0.176\omega^2.\tag{3.43c}$$

where ω is Pitzer acentric factor. The Peng-Robinson (1976) form of the equation is

$$Z = \frac{V}{V - b} - \frac{aV}{RT[V(V + b) + b(V - b)]},\tag{3.44}$$

with

$$a = 0.45724 \frac{R^2 T_c^2}{P_c} \alpha(T_r, \omega),\tag{3.45a}$$

$$b = 0.0778 \frac{RT_c}{P_c},\tag{3.45b}$$

$$\alpha(T_r, \omega) = [1 + x(1 - \sqrt{T_r})]^2,\tag{3.45c}$$

$$x = 0.37464 + 1.54226\omega - 0.26992\omega^2.\tag{3.45d}$$

Note that the mixing rules given by equations (3.40) and (3.41) are usually used for all the modified Redlich-Kwong EOSs, with the binary interaction coefficients k_{ij} determined by fitting experimental data on binary systems.

The various Redlich-Kwong-type equations are cubic in that they can be rewritten in a form

$$Z^3 + B_1 Z^2 + B_2 Z + B_3 = 0,\tag{3.46}$$

where the B_i coefficients are functions of temperature, pressure, and the a and b parameters in the particular EOS.

The EOSs provide an accurate method of estimating fluid densities from temperature, pressure, and composition data and, as such, have been used for this purpose in conjunction with other methods of estimating phase equilibriums (e.g., convergence pressure methods) (Williams et al. 1980). However, by applying various thermodynamic relationships to an analytical EOS, the EOS can be used in determining thermodynamic properties such as the fugacity, Gibbs's free energy, enthalpy, and entropy for a mixture and hence can be used directly in determining phase equilibriums (Reid et al. 1977). The use of an EOS in computing phase equilibriums is summarized in the following.

At equilibrium, the chemical potentials for each component must be the same in all phases. This condition can be expressed in terms of fugacity for a two-phase vapor-liquid system as

$$f_i^V = f_i^L, \tag{3.47}$$

where

f = fugacity,
V = vapor,
L = liquid.

Note that Baker et al. (1981) point out that in addition to equality of fugacities, a minimum in Gibbs's free energy is also required at equilibrium.

Fugacity is related to the temperature, pressure, and composition of a mixture by

$$f_i^V = \psi_i y_i P$$

and

$$f_i^L = \psi_i x_i P,$$

where ψ_i is the fugacity coefficient of component i and is given by (Reid et al. 1977)

$$RT \ln \psi_i = \int_{-\infty}^{V} \left(\frac{\delta P}{\delta N_i} - \frac{RT}{V} \right) dV - RT \ln(Z), \tag{3.48}$$

where V is the total volume of a mixture of N moles and N_i is the moles of component i in the mixture.

An analytic EOS provides the necessary information to evaluate equation (3.48). Using equation (3.38) in equation (3.48) gives, for the vapor phase,

$$f_i^V = \psi_i P y_i = \frac{P y_i}{Z^V - B} \left(\frac{Z^V + B}{B} \right)^{E_i} \exp \left[\frac{B_i}{B} (Z^V - 1) \right], \tag{3.49}$$

with

$$E_i = \frac{A}{B} \left(\frac{B_i}{B} - \frac{Z}{A} \sum_{j=1}^{N} y_j A_{ij} \right), \tag{3.50}$$

$$A_{ij} = \frac{a_{ij}P}{R^2 T^{2.5}}, \qquad (3.51)$$

$$B_i = \frac{b_i P}{RT}, \qquad (3.52)$$

and a_{ij} is given by equation (3.41) and Z^V is the compressibility of the vapor phase. An equation similar to (3.49) can be written for each of the phases present and an iterative technique used to determine the individual phase compositions that give equality of fugacities for all components.

Determining the number of phases present at a given temperature and pressure is not a trivial matter, particularly in systems where more than two phases may coexist. Baker et al. (1981) and Fussell (1979) discuss methods for determining the number of phases present utilizing EOSs.

3.10   **CARBON DIOXIDE–CRUDE OIL SYSTEMS: TYPICAL LABORATORY EXPERI-MENTS.** Previous sections have discussed the multiple mechanisms that may occur during a CO_2–crude oil displacement. These mechanisms and their tie to complex phase behavior are dependent on crude composition, reservoir temperature, displacement pressure, and a reservoir's petrophysical makeup.

Before the reservoir engineer can make sophisticated predictions concerning a given CO_2 flood's performance, a suite of experiments must be performed to help quantify the phase behavior and fluid properties at reservoir conditions. Typical data that might be determined include

minimum miscibility pressure,
crude oil swelling and viscosity reduction,
single- and multiple-contact phase volumes and composition, and
asphaltene precipitation.

Laboratory tests to determine this information most often fall into three general categories: (1) high pressure volumetric (PVT) and vapor-liquid equilibrium (VLE) experiments, (2) slim tube displacements, and (3) core displacements. These three types of experiments and the information they reveal are thoroughly discussed by Orr et al. (1982). A brief examination of typical tests and their importance is now in order.

3.10.1   **Constant Composition Expansion (or Flash Test).** The procedure for this test involves the expansion of a reservoir fluid in a series of discrete pressure increments from a pressure in excess of the initial reservoir pressure to some pressure usually much lower than the saturation pressure at a constant temperature equal to that of the reservoir. Equilibrium

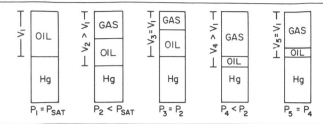

Figure 3.53 Constant composition expansion.

at each pressure level is assured by agitating the cell at each pressure level for which two hydrocarbon phases exist. The lab procedure is illustrated schematically on figure 3.53 (Hg means mercury in the cell).

These data are particularly important for the matching of

saturation pressure,
oil or liquid compressibility at pressures in excess of saturation pressure,
the compressibility of the liquid phase as lighter hydrocarbon components are liberated, and
the compressibility factor of the liberated gas.

One additional piece of information that can be obtained from such a test is the saturated fluid density. Often, the specific gravity of the saturated fluid is reported instead and is usually a calculated value obtained from the fluid composition and individual component gravities. This is not accurate enough for tuning the correlations used in simulation, and therefore, the saturated fluid density always should be measured by the lab.

It is also important to recognize the fact that such single-contact, constant-composition testing does not model those phases that actually occur during the displacement process. An infinite number of CO_2-oil mixtures will be developed in the reservoir, making it almost impossible to relate displacement efficiency to single-contact information.

More recent tests by Gardner et al. (1981) and Orr et al. (1982) have refined the traditional PVT analysis by introducing stepwise multiple-contact experiments that mix either the CO_2-rich phase from a binary mixture with fresh reservoir fluid or the oil-rich phase with fresh CO_2 or continuous multiple-contact experiments where CO_2 is continuously contacting the reservoir crude.

3.10.2 **Constant Volume Depletion.** The constant volume depletion experiment is similar to the flash liberation test in that it involves stepwise pressure reductions in the fluid cell. However, this test starts at the saturation pressure of the fluid and progresses through five or six pressure decrements while enough gas is withdrawn at each pressure level to restore the cell to its initial volume. The lab procedure is illustrated schematically in figure 3.54.

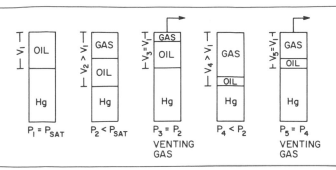

Figure 3.54 Constant volume depletion.

The following types of data can be obtained from this type of test:

the saturation pressure (either bubble point or dew point),
the compressibility of the gas removed at each step,
the liquid saturation as a fraction of the cell volume,
the composition of the gas removed at each step, and
the liquid and vapor viscosities, the liquid usually measured in parallel experiments
    and the gas usually calculated from gas composition.

This type of test is somewhat less common than the constant composition (i.e., flash) or differential liberation tests, which are conducted on almost all fluid samples. Nevertheless, if a compositional study is anticipated, the data from this experiment are particularly important in the characterization of

the compressibility factor of the gas liberated as it varies with composition,
the equilibrium K-values of the components that make up the fluid sample through
    duplicating the composition of the gas removed from the cell, and
the compressibility of the liquid phase.

Since this laboratory test provides compositional analysis of a fluid phase during pressure depletion, the need for reliable data is paramount. The lab should be advised beforehand that all possible care should be taken during the experiment especially in both reading volumes and determining a representative molecular weight of any heavy pseudocomponent (i.e., C_7+). The molar balance of initial fluid versus fluids removed is critical to the estimation of K-values.

3.10.3   **Differential Liberation.** The differential liberation experiment is similar to the constant volume depletion technique since it involves pressure reduction from the saturation pressure and gas removal after equilibration at each pressure level. However, the differential liberation is performed on oil samples only, and at each pressure level, all the evolved gas is removed regardless of the resulting cell volume in relation to its initial volume.

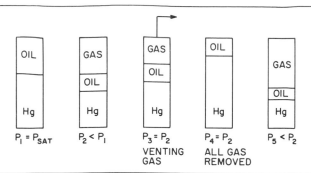

Figure 3.55 Differential liberation.

The lab procedure is depicted schematically in figure 3.55.

The following lab data can be obtained during a test of this type:

oil formation volume factor reported as a fraction of the residual oil volume,
gas volume liberated during any pressure decrement that can be converted to
  dissolved gas ratio,
oil density,
gas compressibility factor,
gas gravity, and
oil viscosity, which is usually measured in a parallel experiment.

The differential liberation is the most common laboratory test and is almost always performed on reservoir oil samples. The data may be used in tuning the fluid property correlations for the following characteristics:

reservoir oil bubble point;
dissolved gas ratio, which gives an estimate of the relationship of equilibrium K-
  values with pressure;
oil density as a function of pressure, which is related to oil compressibility;
gas compressibility factor as it varies with pressure; and
gas gravity or density, which is related to the equilibrium K-values.

The reported oil densities from this type of test may be calculated values derived from a calculated saturated density and the measured relative volume. The lab should be advised that actual measured densities are preferred. Often, the reported viscosities are obtained from a parallel experiment with a similar oil. If this is true, it should be noted in the report since a laboratory-defined similar oil may be significantly different in composition, therefore having different viscosity characteristics.

3.10.4 **Separator Tests.** The separator test is comparable to the constant composition except that only one pressure decrement is used. The lab procedure is to measure a volume of reservoir oil at reservoir temperature and saturation pressure, then to cool it to the projected separator temperature and decrease the pressure in a single stage to the projected separator pressure. The data obtained from such a test include:

gas evolved from reservoir to surface separator conditions and from separator to stock tank conditions,
the shrinkage of reservoir oil to surface conditions as a result of gas evolution, and
the density of residual oil and gravity of liberated gas.

In the absence of constant volume depletion data, these data may be used to obtain an estimate of component equilibrium K-values that affect the gas evolution rates. However, care should be taken in using these data since most EOSs and fluid property correlations are not accurate at low pressures (i.e., less than 200 psi). The primary contribution of separator tests is in assisting operations personnel in defining the set of surface separation conditions that will maximize stock tank oil production in relation to reservoir withdrawals.

3.10.5   **Description of Special Laboratory Tests.** In addition to the previously described lab tests, which are fairly standard, a number of other more exotic tests may be performed for very specific applications. If a reservoir is to be depleted under a gas injection, miscible (hydrocarbon or otherwise), or dry-gas-cycling scheme, any one or a number of these tests should be performed.

SWELLING TEST. During a swelling test (fig. 3.56), gas of known composition (usually similar to proposed injection gas) is added to reservoir oil in a series of steps. After each gas addition, the cell is pressured up until only one phase is present. The gas addition starts at the bubble point of the reservoir fluid sample and continues to perhaps 80 mole percent injected gas in the fluid sample.
The data obtained from such a test include:

the relationship of saturation pressure with volume of gas injected;
the saturation pressure may change from a bubble point to a dew point after significant volumes of gas injection; and
the volume of the saturated fluid mixture in relation to the volume of the original saturated reservoir oil.

These data may be used to characterize the mixing of the individual hydrocarbon components and the effect of mixing on the volume increase of the saturated fluid and the ability of the hydrocarbon mixture to dissolve injection gas.
Before conducting such a test, the lab should be advised that reliability of volume measurements is critical for future work and that all care should be taken in obtaining accurate measurements. Often, the cross-sectional area of the cell is so large that a change of 10% in the total fluid volume cannot be read with accuracy; that is, if the volume can be read to within 1% accuracy and the swelling of the fluid is 10% of the total volume, then in fact, the accuracy of the reading is only within 10% of the volume

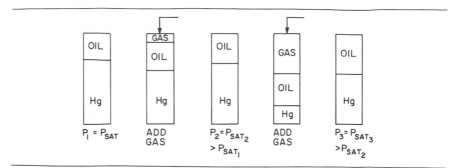

Figure 3.56 Swelling test.

change. This is not accurate enough for application to fluid properties matching. In addition, the lab should be advised if a solid phase is expected to precipitate since such a phase may be interpreted as a liquid, thereby resulting in unreliable saturation pressures.

SLIM TUBE DISPLACEMENT. Unlike the previously described laboratory tests that concentrate specifically on fluid property variations with pressure and composition, the slim tube test (fig. 3.57) is conducted to examine flushing efficiency and fluid mixing during a miscible displacement process. Its purpose is solely to examine the phase behavior properties for a given CO_2 displacement by eliminating reservoir heterogeneities, water, and gravity.

This type of experiment is the most commonly performed to evaluate operating pressures for a potential CO_2 flood. Even so, experimental techniques have differed substantially as Orr et al. (1982) report in table 3.3. Sample results from one such slim tube test are shown in figure 3.58. As expected, oil recovery increased sharply from a low pressure, immiscible displacement to a high pressure, vaporizing, condensing, or first-contact miscible displacement with the miscibility pressure easily located.

The laboratory test is performed in a long tube (i.e., 15 m or longer) with a relatively small diameter (i.e., less than 1 cm) packed with glass beads or sand. This apparatus is saturated with reservoir oil and pressured up to the anticipated pressure level of the miscible flood scheme. The reservoir oil is then displaced, at a regulated pressure, with the fluid proposed for the miscible flood. Volumes of produced fluids are recorded as functions of the number of pore volumes of fluid injected. Some component in the injection stream is selected as a tracer, and its concentration in the produced stream is recorded as a function of the pore volumes of fluid injected. This enables detection of the displacement front at the exit end of the slim tube.

The results of such experiments are useful in determining whether a simulator can provide a reasonable approximation to the miscible displacement process as it occurs at actual reservoir pressure and temperature conditions. By varying a number of characteristics of the computer model (i.e., relative permeability curves, time step size), the user can get a general impression of their effects on predicted flood performance. Making a series of runs with a range of grid increments will provide the user with some insight as to the magnitude of numerical dispersion effects.

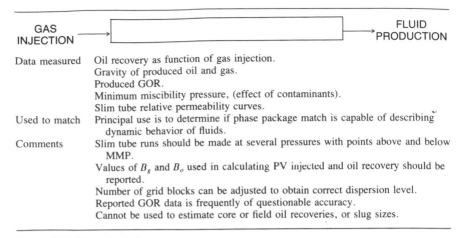

Figure 3.57 Slim tube test.

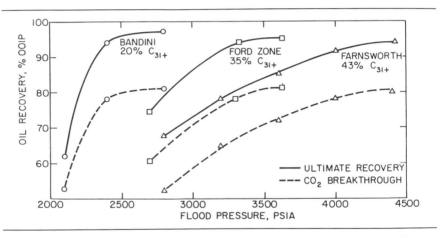

Figure 3.58 Determination of miscibility using CO_2 at 165°F (Holm and Josendal 1982) ©SPE-AIME.

CORE DISPLACEMENT. The core displacement test (fig. 3.59) is virtually identical in procedure to the slim tube displacement test except that a reservoir core sample replaces the bead-packed tube. The method of conducting the test and the results monitored are identical to the slim tube test. Similarly, the simulation procedure and sensitivity investigations made are identical. However, CO_2 core floods are most difficult to interpret since linear displacements may or may not exhibit characteristics common to a given reservoir such as dispersion, mobile water, wettability, viscous fingering, gravity segregation, oil bypassing due to heterogeneity, and trapped oil saturation. Gardner and Ypma (1982) have investigated CO_2 core flood scaling, while Tiffin and Yellig (1982) have reviewed the effect or noneffect of mobile water on oil recovery. Other recent laboratory core flood work has been listed by Orr et al. (1982) in table 3.4.

Table 3.3 Characteristics of Slim Tube Displacement Experiments

Author(s)	Length (m)	Inside Diameter (cm)	Geometry	Packing Material	Permeability (darcies)	Porosity (%)	Rate (cm/h)[a]
Rutherford (1962)	1.5	0.98	Vertical tube	50–70 mesh Ottawa sand	24	35	37
Yarborough and Smith (1970)	6.7	0.46	Flat coil	No. 16 AGS Ottawa sand (140–200 mesh)	2.74	36	66
Holm and Josendal (1974)	14.6	0.59	Spiral coil	No. 60 Crystal sand	20	39	381
Holm and Josendal (1982)	15.8	0.59	Spiral coil	No. 60 Crystal sand	20	39	101 to 254
Huang and Tracht (1974)	6.1	1.65	Spiral coil	Crushed Berea sandstone	1.78	43	4.7
Yellig and Metcalfe (1980)	12.2	0.64 OD	Flat coil	160–200 mesh sand	2.5	42	51 to 102
Peterson (1978)	17.1	0.64	Spiral oval	60–65 mesh sard	19	35	48
Wang and Locke (1980)	19.3	0.62	Spiral coil	80–100 mesh glass beads	13	35	381
Orr and Taber (1981)	12.2	0.64	Spiral coil	170–200 mesh glass beads	5.8	38	41
Gardner et al. (1981)	6.1	0.46	Flat coil	230–270 mesh glass beads	1.4	37	32 / 64
Sigmund et al. (1979)	17.9	0.78	Spiral coil	140 mesh glass beads	5	42	81
Johnson and Pollin (1981)	15	0.64	Flat coil	100–140 mesh glass beads	14	34	500

Sources: J. P. Johnson and J. S. Pollin, "Measurements and Correlation of CO_2 Miscibility Pressures," SPE, Tulsa (1981); Orr et al., "Laboratory Experiments to Evaluate Field Prospects for CO_2 Flooding," *J. Pet. Tech.* (April 1982).
[a]1 cm/h = 0.787 ft/D.

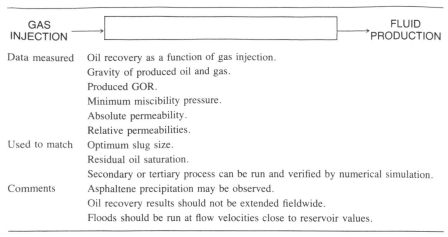

GAS INJECTION →		→ FLUID PRODUCTION

Data measured   Oil recovery as a function of gas injection.
Gravity of produced oil and gas.
Produced GOR.
Minimum miscibility pressure.
Absolute permeability.
Relative permeabilities.
Used to match   Optimum slug size.
Residual oil saturation.
Secondary or tertiary process can be run and verified by numerical simulation.
Comments   Asphaltene precipitation may be observed.
Oil recovery results should not be extended fieldwide.
Floods should be run at flow velocities close to reservoir values.

Figure 3.59 Core floods.

However, even with their drawbacks, Orr et al. (1982) feel core floods do seem appropriate for answering at least three questions directly: (1) Can CO_2 mobilize tertiary oil under conditions that are closer to field displacement conditions than those occurring in slim tube displacements? (2) What is the residual saturation in the swept zone of a CO_2 displacement? (3) Does CO_2 injection alter reservoir permeability by asphaltene precipitation or dissolution of minerals?

Core floods may also be important for testing numerical simulation methods. The use of core displacement tests in validating a computer simulator is, however, extremely limited. This is due to a number of drawbacks to using a core sample as opposed to a slim tube—for example,

Core samples, no matter how small they are, are seldom, if ever, homogeneous. Any minute heterogeneities cannot be simulated with a one-dimensional model.
For a horizontal flood, horizontal cores are required with the longest possible being less than four inches (i.e., the diameter of a full four-inch core). Thus, tacking a series of smaller diameter cores end to end is necessary. This results in discontinuous pore spaces and poor scaling at the contacting faces.
In full core radial samples, it is extremely difficult if not impossible to get a complete seal between core and core holder; therefore, bypassing of injected fluids is a problem.

For these reasons, reported lab data can be contradictory or at best inconsistent. Acceptance of these data can lead to erroneous conclusions from any core flood simulation. Therefore, use core flood lab results with extreme caution.

3.11   UNCERTAINTIES IN LABORATORY DATA. Fluid properties data, as obtained from laboratory tests, are often at best a reasonable approximation to the properties exhibited by the fluid system in the reservoir. On the one hand, this is primarily a result of poor

Table 3.4 Characteristics of CO_2 Displacements in Cores

Author(s)	Length (m)	Diameter (cm)	Rock Type	Permeability (md)	Porosity (%)	Velocity (cm/h)[a]	Core Orientation	Secondary (s), Tertiary (t)
Rathmell et al. (1971)	1.8, 6.1, 13	5.1	Boise outcrop sandstone	1,000	27	12.7 to 25.4	Vertical, horizontal	s
Shelton and Schneider (1975)	0.3, 1.2, 2.4	5.1	Berea sandstone	135 to 890	16 to 22	0.3 to 40.6	Horizontal	s, t
Shelton and Yarborough (1977)	4.9	5.1	Berea sandstone	688	20	3.8	Horizontal	s
Holm (1963)	0.15, 0.3, 2.3	8.9	Berea sandstone, McCook dolomite	150 to 250, 100 to 140	14 to 21, 16 to 20	6.3 to 31.2	Horizontal	s
Sigmund et al. (1979)	0.4	3.2	Butted reservoir cores	76	15	4.6	Horizontal	s, t
Huang and Tracht (1974)	1.8	5.1	Berea sandstone	440 to 470	24	2.5	Horizontal	t
Rosman and Zana (1977)	0.8	2.5	Butted reservoir cores	11	15	3.6	Horizontal	s
Graue and Zana (1972)	0.9	2.5	Rangely reservoir cores	59	18	1.3	Horizontal	t
Metcalfe and Yarborough (1979)	2.4	5.1	Berea sandstone	450		2.7	Horizontal	s
Watkins (1978)	0.3	3.5 to 5.1	Berea sandstone	290 to 670	17 to 22	0.4 to 19.1	Vertical	s, t
	0.15	5.1 to 6.4	San Andres carbonate reservoir cores	6.2 to 16.8	11 to 15	0.3 to 5.1	Vertical	s, t
Yellig (1981)	2.4, 4.9	5.1	Berea sandstone	113 to 826	19 to 21	0.55 to 17.7	Horizontal	s
Doscher and El-Arabi (1981)	1	5.7	Sandstone	3,000	—	0.06 to 0.5	Horizontal	s, t
Gardner and Ypma (1982)	0.15, 0.23, 0.47, 0.57	7.6, 5.1, 5.1, 2.5	Berea sandstone	70	—	1	Vertical	s
Tiffin and Yellig (1982)	2.4	5.1	Berea sandstone	—	21	1.8 to 4.2	Horizontal	s, t

Source: Orr et al., "Laboratory Experiments to Evaluate Field Prospects for CO_2 Flooding," *J. Pet. Tech.* (April 1982).
[a] 1 cm/h = 0.787 ft/D.

control (i.e., nonrepresentative sampling) and inherent experimental inaccuracies. On the other hand, the compositional simulator is a far more precise tool in generating fluid properties from the composition and component characterization. It is important that the reservoir planner recognize the possible inaccuracies of lab data and the precision of the simulation tool prior to conducting a compositional study so the study may proceed efficiently. For example, attempting to tune the bubble point correlations to within 5 psi of observed values is not justifiable when the lab value is accurate to within ± 10 psi and the sample is representative to within ± 50 psi. The additional efforts required to obtain the better calculated values cannot be justified costwise and probably will not provide any better results from the reservoir studies.

Some of the uncertainties and errors that can occur are summarized in the following sections with precautions that may be taken to prevent them.

3.11.1   Reservoir Fluid Sampling. Ideally, the most representative fluid samples will be obtained by bottomhole sampling when the flowing bottomhole pressure is greater than the fluid saturation pressure. If the flowing bottomhole pressure is less than the saturation pressure, liberated gas may escape the collector vessel, resulting in nonrepresentative sampling. In this event, collection of surface separator gas and oil followed by recombination in the proper ratio will probably be more representative. However, this type of sampling assumes that the fluids produced in the wellbore are representative of the true reservoir fluid (i.e., no release of solution gas in the reservoir) and that the producing separator gas-oil ratio is reasonably consistent with time. The samples subject to the greatest degrees of uncertainty would be those collected either downhole or at the surface a long time after the reservoir has passed through its saturation pressure. If at all possible, these samples should be avoided.

3.11.2   Sample Consistency. Often, samples collected at approximately the same time from different wells in the same reservoir can exhibit significantly different properties. As an example, bubble points can vary from 900 psi to 1,300 psi. Aside from measurement error in the lab, these differences are mainly due to poor sampling techniques and sample contamination. If you are faced with this type of variance, there is probably no way to obtain a good match of all the data and therefore it is necessary either to decide which sample is most representative of the reservoir fluid and match that data or to obtain a match of the most reliable data obtained from all the tests. The first option is probably the most advisable provided that at least one of the samples is reliable.

3.11.3   Laboratory Measurement Accuracy. Volumes reported by the lab that are used in relative volume and density calculations may be subject to varying degrees of error depending on the equipment used and the care taken by the individual running the test. Ideally, the volume measurement apparatus will be precise enough to measure relatively small volume changes in relation to the total sample volume. The only precaution that can be taken here is to plot and smooth the observed volume and density data prior to conducting the computer study of fluid properties. Reported saturation pressures can also be subject to significant error since they depend on the ability of lab personnel in detecting the formation of a second hydrocarbon phase. If the fluid sample volume is beyond the range of the sightglass, the reported value is, at best, a guess. Since laboratory measurement accuracy is primarily a function of the experience and concentration of the lab personnel, all efforts should be made in selecting a reliable laboratory and requesting the services of the best personnel available.

3.11.4 **Resolving Uncertainties.** Before attempting to match laboratory data with the EOS, time should be taken to review and, if necessary, adjust the lab data. All data should be plotted to ensure that successive readings are consistent. Inconsistent values should be either discarded or corrected. The reservoir planner should contact the lab and obtain some estimate of the experimental error associated with each type of measurement. This will ensure that time is not wasted refining a match beyond the accuracy of the data.

3.12 **DATA REQUIREMENTS FOR MISCIBLE GAS INJECTION MODELING.** In a miscible gas injection process, like CO_2, the injected fluid is usually not miscible with the reservoir oil in the classical sense of being capable of being mixed in all proportions and still remain as a single phase. Rather, as the injected gas is added stepwise to the reservoir oil at constant temperature and pressure, it will dissolve in the oil up to the point at which the oil becomes saturated with the gas. Any further addition of gas will result in the formation of a second phase (either a vapor or liquid phase depending on whether the pressure is below or above the pseudo–critical pressure of the system) and perhaps even a third (multiple liquid) and fourth (asphaltene) phase.

While these gases are not truly miscible with the reservoir oil, they still are capable of achieving oil recoveries approaching those of truly miscible fluids. These high recoveries are the result of mass transfer between the equilibrium phases. In certain instances, the mass transfer results in an enriched oil phase whose composition changes with time and sooner or later becomes miscible with the injected CO_2. This process, as has been described earlier, is frequently termed a condensing, multiple-contact miscible displacement. In other cases, the injected gas is so efficient at vaporizing the hydrocarbons from the liquids that high recoveries can be obtained even when the original gas composition does not approach a miscible composition. The interphase mass transfer can also lead to gas and liquid compositions that have very low interfacial tensions so that high recoveries are achieved by a reduction in the residual liquid saturation.

In order to model these processes, it is necessary to couple the fluid flow description with a method of analytically representing phase equilibrium and phase properties. The approach used is to develop a reasonable number of pseudocomponents (four to eight) and their corresponding EOS parameters that model experimentally determined phase

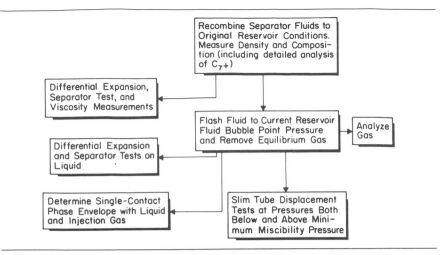

Figure 3.60 Outline for experimental procedures for obtaining fluid data necessary for simulation of a *miscible* CO$_2$ displacement process.

behavior data. The experiments required to achieve this are outlined in figure 3.60. The first step is to determine the properties of the reservoir fluid both at its original conditions and at the conditions at which the gas injection recovery process is to be conducted. Then, the phase behavior of the injection gas and reservoir fluid is determined in a set of experiments similar to a standard swelling test, where injection gas is added to the reservoir fluid in increments and the saturation pressure and density of the saturated fluid is measured. Experience has shown that it is necessary to continue the gas additions beyond the critical composition of the system (i.e., the point where saturation pressure changes from a bubble point to a dew point) to quite high CO$_2$ mole fractions (90% +) in order to characterize accurately the phase behavior of the system. In addition, for each mixture, the volume of liquid and vapor should be measured at several pressures below the mixture's saturation pressure. These data are critical to determining the ability of the injection gas to vaporize the oil. Figure 3.61 shows a typical phase envelope (P-X diagram) determined from such a set of experiments for CO$_2$. Also note on this figure the prediction values of the Peng-Robinson EOS when attempting to match the laboratory results.

Some of the preceding measurements (swelling tests, bubble point, etc.) are normally performed in a cell with a quartz window. The cell should be observed visually to determine if any asphaltic particles have precipitated after contact with the injection gas. In addition, any appearance of multiple phases should be reported and their volumes measured.

Slim tube displacement experiments should be run using the injection gas and crude oil at several pressures to determine the effect of pressure on the ability of the gas to recover the oil in a dynamic situation. The pressure levels used should extend from well below the minimum miscibility pressure to pressures well above this point.

In summary, then, the minimum amount of data necessary to describe the CO$_2$–crude oil displacement process adequately is given in the following sections.

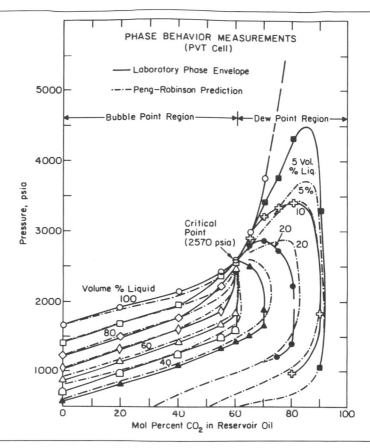

Figure 3.61 Pressure-composition diagram for a typical CO₂-oil system at 130°F (Hong 1982) ©SPE-AIME.

3.12.1 **Properties of Crude Oil.** The following data should be available for both the original reservoir fluid and for a sample recombined to current reservoir conditions:

analysis of sample including extended analysis of C_7+ fraction;

formation volume factor as function of pressure, as measured in a differential expansion (DE);

solution gas-oil ratio as function of pressure as measured in a DE; gas gravity and compressibility (Z-factor) should be measured;

viscosity of saturated liquid at several pressures.

3.12.2 **Carbon Dioxide–Crude Oil Properties for Immiscible Carbon Dioxide Projects.** ''Immiscible CO_2 project'' implies that CO_2 can dissolve into the oil to a certain extent (determined by pressure), but when a gas phase forms, it will contain negligible hydrocarbons (i.e., no stripping effect of CO_2). For this case, the principal recovery mechanisms are liquid swelling and viscosity reduction:

swelling test data giving bubble point pressures and swollen volumes up to the maximum pressures to be encountered in the reservoir,
liquid viscosities as a function of pressure and CO_2 content.

3.12.3   Carbon Dioxide–Crude Oil Properties for Miscible Carbon Dioxide Projects. "Miscible project" implies that, when two phases exist in the reservoir, significant amounts of the hydrocarbon components will exist in both phases, giving rise to the possibility of a multiple-contact miscible displacement:

Swelling test data giving saturation pressure and swollen volumes. The CO_2 additions should be continued to well above the critical composition of the CO_2–crude oil system (i.e., both bubble points and dew points should be observed). At each CO_2 addition, liquid saturation should be measured for pressures below the saturation pressure.
Liquid viscosities should be measured as a function of pressure at a few CO_2 concentrations.
Slim tube displacements should be conducted at several pressures both below and above the minimum miscibility pressures.

3.13   PSEUDOCOMPONENTS. While the application of EOSs to simple mixtures is relatively straightforward, crude oil systems pose many seemingly insurmountable problems. The essentially infinite number of components contained in a typical crude oil makes it impossible to obtain a complete chemical analysis. Standard crude oil analysis lumps all components heavier than hexane into a single C_7^+ pseudocomponent characterized only by its average molecular weight and density. Extended analyses only give a boiling point distribution of the C_7^+ fraction and possible amounts of naphenic, paraffinic, and aromatic compounds. Even if it were possible to obtain a full analysis of a crude oil by component, the critical properties for many of the components would not be known. In addition, the information needed to determine the binary interaction coefficients is available for only a relatively few compounds. Therefore, the application of an EOS model to the calculation of crude oil systems by necessity requires the use of pseudocomponents.

In the *first-contact miscible* case, one is not concerned with saturation pressure (bubble point) or flash calculations; therefore, the EOS need only be used to calculate density and viscosity (using the Lohrenz et al. (1964) method and EOS density) of single-phase solvent-oil mixtures. Ideally, one would obtain parameters for two pseudocomponents such that the EOS for these two pseudocomponents will give the same density and viscosity versus composition and pressure as for the eight-component system.

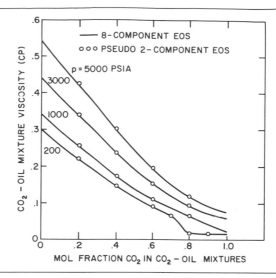

Figure 3.62 Viscosity versus pressure and composition for CO₂-oil mixtures (Paul et al. 1980).

Figures 3.62 and 3.63 show viscosity and density for CO₂–crude oil mixtures versus pressure and composition calculated from eight-component and pseudo-two-component EOS calculations. The pseudo-two-component calculations almost duplicate the eight-component EOS results (Paul et al. 1980).

For *multiple-contact miscible* cases where two or three phases are present, composition dependent oil and gas viscosities, interfacial tensions, and relative permeabilities for changing interfacial tensions must also be included with the EOS package prior to its coupling with a compositional simulator. Gas and oil densities and compositions are, of course, predicted by the EOS. Viscosities of oil and gas can be computed from the Jossi et al. (1962) correlation. The gas/oil interfacial tension and its effect on relative permeabilities can be calculated from the Macleod-Sugden (see Reid et al. 1977), Coats (1980), and Ngiem et al. (1981) models.

Unlike the first-contact miscible case, it may be difficult to accurately match laboratory-observed phase behavior if the process is multiple-contact miscible or the phase behavior includes the appearance of multiple liquids. Fussell (1979), for example, needed over 20 components in a modified Redlich-Kwong EOS to predict this type of phase behavior accurately. The number of pseudocomponents required and their characterization are also discussed by Katz and Firoozabadi (1978) and by Williams et al. (1980) for hydrocarbon systems and by Turek et al. (1980) for CO₂–crude oil systems. These authors used from 12 to as many as 40 components to characterize a crude oil. Since the cost of running a compositional reservoir simulator (see chapter 4) is a strong function of the number of components being used, the use of this many components would make the modeling of anything but small laboratory experiments prohibitively expensive.

While it may require 12 to 40 components to predict (without parameter modification) CO₂-crude oil equilibria accurately, Hong (1982) has developed a method of pseudoiza-

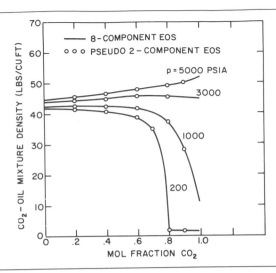

Figure 3.63 Density versus pressure and composition for CO_2-oil mixtures (Paul et al. 1980).

tion that systematically adjusts C_7^+ critical values and the interaction coefficients. His method allows the phase equilibriums of a simple CO_2 system (no multiple liquids) to be modeled with as few as 4 to 8 components without a significant loss in accuracy in describing the phase behavior and, hence, a major reduction in computing time. If successful pseudoization of components can be achieved, a major obstacle to modeling the CO_2 process will be removed.

REFERENCES

Alston, R. B., Kokolis, G. P., and James, C. F.: "CO_2 Minimum Miscibility Pressure: A Correlation for Impure CO_2 Streams and Live Oil Systems," paper SPE 11959 presented at the Fifty-Eighth Annual Technical Conference, San Francisco, Oct. 5–8, 1983.

Aziz, K., and Settari, A.: *Petroleum Reservoir Simulation,* Applied Sciences Publishers, London (1979).

Baker, L. E., Pierce, A. C., and Luks, K. D.: "Gibbs Energy Analysis of Phase Equilibria," paper SPE 9806, SPE/DOE 2nd Joint Symposium on Enhanced Oil Recovery of the Society of Petroleum Engineers, Tulsa, April 5–8, 1981.

Bardon, C., and Longeron, D. G.: "Influence of Very Low Interfacial Tensions on Relative Permeability," *Soc. Pet. Eng. J.* (Oct. 1980) 391–401.

Beal, C.: "The Viscosity of Air, Water, Natural Gas, Crude Oil, and Its Associated Gases at Oil Field Temperatures and Pressures," *Trans., AIME* (1946) **165,** 94–115.

Beggs, H. D., and Robinson, J. R.: "Estimating the Viscosity of Crude Oil Systems," *J. Pet. Tech.* (Sept. 1975) 1140–1141.

Bridgeman, D. W.: *The Physics of High Pressure,* Macmillan Press, New York (1931).

Burcik, E. J.: *Petroleum Reservoir Fluids,* IHRDC, Boston (1956).

Canjar, L. N., and Manning, F. S.: *Thermodynamic Properties and Reduced Correlations for Gases,* Gulf Publishing, Houston (1967).

Carr, N. L., Kobayashi, R., and Burrows, D. B.: "Viscosity of Hydrocarbon Gases under Pressure," *Trans., AIME* (1954) **201,** 264–272.

Chew, J., and Connally, C. A., Jr.: "A Viscosity Correlation for Gas-Saturated Crude Oils," *Trans., AIME* (1959) **216,** 23–25.

Coats, K. H.: "An Equation of State Compositional Model," *Soc. Pet. Eng. J.* (Oct. 1980) 363–376.

Craft, B. C., and Hawkins, M. F.: *Applied Petroleum Reservoir Engineering,* Prentice-Hall, Englewood Cliffs, N.J. (1959).

Cragoe, C. E.: *Thermodynamic Properties of Petroleum Products,* Monograph 10 Bureau of Mines, Washington, D.C. (1957) **1.**

Crawford, H. R., Neill, G. H., Bucy, B. J., and Crawford, P. B.: "Carbon Dioxide—A Multipurpose Additive for Effective Well Stimulation," *J. Pet. Tech.* (March 1963) 237–242.

Cronquist, C.: "Carbon Dioxide Dynamic Miscibility with Light Reservoir Oils," presented at the 4th Annual U.S. DOE Symposium, Tulsa, Aug. 28–30, 1978.

Culberson, O. L., and McKetta, J. J.: "Phase Equilibria in Hydrocarbon-Water Systems," *Trans., AIME* (1951) **192,** 223–227.

David, A.: "Asphaltene Flocculation During Solvent Stimulation of Heavy Oils," paper 47d presented at the 71st National Meeting of American Institute of Chemical Engineers, Dallas, Feb. 20–23, 1972.

de Nevers, N.: "Carbonated Waterflooding," *World Oil* (Sept. 1966) 93–96.

Dodds, W. S., Stutzman, L. F., and Sollami, B. J.: "Carbon Dioxide Solubility in Water," *J. Ind. and Eng. Chem.* (1956) **1,** 92–95.

Doscher, T. M., and El-Arabi, M.: "High Pressure Model Studies of Oil Recovery by Carbon Dioxide," paper SPE/DOE 9787 presented at the 2nd Joint SPE/DOE Symposium on Enhanced Oil Recovery, Tulsa, April 5–8, 1981.

Fatt, I.: "Pore Volume Compressibility of Sandstone Reservoir Rocks," *J. Pet. Tech.* (March 1958) 64–66.

Fayers, F. J., Hawes, R. I., and Mathews, J. D.: "Some Aspects of the Potential Application of Surfactants or CO_2 as EOR Processes in North Sea Reservoirs," *J. Pet. Tech.* (Sept. 1981) 1617–1627.

Francis, A. W.: "Ternary Systems of Liquid Carbon Dioxide," *J. Phys. Chem.* (1959) 1099.

Fussell, L. T.: "A Technique for Calculating Phase Equilibria of Three Coexisting Hydrocarbon Phases," *Soc. Pet. Eng. J.* (Aug. 1979) 203–210.

Gardner, J. W., Orr, F. M., and Patel, P. D.: "The Effect of Phase Behavior on CO_2 Flood Displacement Efficiency," *J. Pet. Tech.* (Nov. 1981) 2067–2081.

Gardner, J. W., and Ypma, J. G. J.: "An Investigation of Phase Behavior–Macroscopic Bypassing Interaction in CO_2 Flooding," paper SPE/DOE 10686 presented at the 3rd Joint DOE/SPE Symposium on Enhanced Oil Recovery, Tulsa, April 4–7, 1982.

Goodrich, J. H.: "Review and Analysis of Past and Ongoing Carbon Dioxide Injection Field Tests," paper SPE/DOE 8832 presented at the First Joint DOE/SPE Symposium on Enhanced Oil Recovery, Tulsa, April 20–23, 1980.

Graue, D. J., and Zana, E.: "Study of a Possible CO_2 Flood in the Rangely Field, Colorado," *J. Pet. Tech.* (July 1972) 874–882.

Hall, H. N.: "Compressibility of Reservoir Rocks," *Trans., AIME* (1953) **198,** 309.

Hawkins, G. A., Solberg, H. L., and Potter, A. A.: "The Viscosity of Water and Superheated Steam," *Trans., AIME* (1940) **62,** 677.

Henry, R. L., and Metcalfe, R. S.: "Multiple Phase Generation during CO_2 Flooding," *Soc. Pet. Eng. J.* (Aug. 1983) 595–601.

Holm, L. W.: "CO_2 Requirements in CO_2 Slug and Carbonated Water-Oil Recovery Processes," *Producers Monthly* (Sept. 1963) 6–28.

Holm, L. W., and Josendal, V. A.: "Mechanisms of Oil Displacement by Carbon Dioxide," *J. Pet. Tech.* (Dec. 1974) 1427–1438.

Holm, L. W., and Josendal, V. A.: "Effect of Oil Composition on Miscible-Type Displacement by Carbon Dioxide," *Soc. Pet. Eng. J.* (Feb. 1982) 87–98.

Hong, K. C.: "Lumped-Component Characterization of Crude Oils for Compositional Stimulation," paper SPE/DOE 10691 presented at the 3rd Joint SPE/DOE Symposium on Enhanced Oil Recovery, Tulsa, April 4–7, 1982.

Huang, E. T. S., and Tracht, J. H.: "The Displacement of Residual Oil by Carbon Dioxide," paper SPE 4635 presented at the SPE 3rd Symposium on Improved Oil Recovery, Tulsa, April 1974.

Hutchinson, C. A., Jr., and Braun, P. H.: "Phase Relations of Miscible Displacement in Oil Recovery," *AIChE J.* (1961) 7, 64–72.

Johnson, J. P., and Pollin, J. S.: "Measurements and Correlation of CO_2 Miscibility Pressures," paper SPE 9790 presented at the 2nd Joint SPE/DOE Symposium on Enhanced Oil Recovery, Tulsa, April 5–8, 1981.

Johnson, W. E., MacFarlane, R. M., and Breston, J. N.: "Changes in Physical Properties of Bradford Crude Oil When Contacted with CO_2 and Carbonated Water," *Producers Monthly* (Nov. 1952) 16.

Jossi, J. A., Stiel, L. I., and Thodos, G.: "The Viscosity of Pure Substances in the Dense Gaseous and Liquid Phases," *AIChE J.* (Jan. 1962) 59–63.

Katz, D. L., and Firoozabadi, A.: "Predicting Phase Behavior of Condensate/Crude-Oil Systems Using Methane Interaction Coefficients," *J. Pet. Tech.* (Nov. 1978) 1649–1654.

Klins, M. A. and Farouq Ali, S. M.: "Heavy Oil Production by Carbon Dioxide Injection," *J. Can. Pet. Tech.* (Sept.–Oct. 1982) 59–72.

Latil, M.: *Enhanced Oil Recovery,* Gulf Publishing, Houston (1980).

Leach, M. P., and Yellig, W. F.: "Compositional Model Studies—CO_2 Oil-Displacement Mechanisms," *Soc. Pet. Eng. J.* (Feb. 1981) 89–97.

Lohrenz, J., Bray, B. G., and Clark, C. R.: "Calculating Viscosity of Reservoir Fluids from Their Composition," *J. Pet. Tech.* (Oct. 1964) 1171–1176.

Martin, J. W.: "Additional Oil Production Through Flooding with Carbonated Water," *Producers Monthly* (July 1951) 18–23.

Mehra, R. K., Heidemann, R. A., and Aziz, K.: "Composition of Multiphase Equilibrium for Compositional Simulation," paper SPE 9232 presented at the 55th Annual Fall Technical Conference and Exhibition, Dallas, September 21–24, 1977.

Menzie, D. E., and Nielsen, R. F.: "A Study of the Vaporization of Crude Oil by Carbon Dioxide Repressuring," *J. Pet. Tech.* (Nov. 1963) 1247.

Metcalfe, R. S.: "Effects of Impurities on Minimum Miscibility Pressures and Minimum Enrichment Levels for CO_2 and Rich-Gas Displacements," *Soc. Pet. Eng. J.* (April 1982) 219–225.

Metcalfe, R. S., and Yarborough, L.: "The Effect of Phase Equilibria on the CO_2 Displacement Mechanism," *Soc. Pet. Eng. J.* (Aug. 1979) 242–252.

Miller, J. A., and Jones, R. A.: "A Laboratory Study to Determine Physical Characteristics of Heavy Oil after CO_2 Saturation," paper SPE 9789 presented at the 2nd Joint SPE/DOE Symposium on Enhanced Oil Recovery, Tulsa, April 5–8, 1981.

National Petroleum Council: *Enhanced Oil Recovery,* Washington, D.C. (1976).

Ngiem, L. X., Fong, D. K., and Aziz, K.: "Compositional Modeling with an Equation of State," *Soc. Pet. Eng. J.* (Dec. 1981) 687–698.

Nolen, J. S.: "Numerical Simulation of Compositional Phenomena in Petroleum Reservoirs," paper SPE 4274 presented at the 3rd SPE Symposium on Numerical Simulation of Reservoir Performance, Houston, January 11–12, 1973.

Oellrich, L., Plocker, U., Prausnitz, J. M., and Knapp, H.: "Equation-of-State Methods for Computing Phase Equilibria and Enthalpies," *Int. Chem. Eng.* (Jan. 1981) 21.

Orr, F. M., and Lien, C. L.: "Phase Behavior of CO_2 and Crude Oil in Low Temperature Reservoirs," paper SPE 8813 presented at the 1st Joint SPE/DOE Symposium on Enhanced Oil Recovery, Tulsa, April 20–23, 1980.

Orr, F. M., Jr., Silva, M. K., and Lien, C. L.: "Equilibrium Phase Compositions of CO_2/Crude Oil Mixtures—Part 2: Comparison of Continuous Multiple-Contact and Slim-Tube Displacement Tests," *Soc. Pet. Eng. J.* (April 1983) 281–291.

Orr, F. M., Jr., Silva, M. K., Lien, C. L., and Pettetier, M. T.: "Laboratory Experiments to Evaluate Field Prospects for CO_2 Flooding," *J. Pet. Tech.* (April 1982) 888–898.

Orr, F. M., Jr., and Taber, J. J.: "Displacement of Oil by Carbon Dioxide," Report DOE/ET/12082-9, Final Report for U.S. DOE, Contract No. DE-AC21-78ET12082 (May 1982).

Parkinson, W. J., and de Nevers, N.: "Partial Molal Volume of Carbon Dioxide in Water Solutions," *J. Ind. Eng. Chem.* (Nov. 1969) 709–713.

Paul, G. W., Ramesh, B., and Gould, T. L.: *Advanced Reservoir Engineering,* Intercomp, Houston (1980).

Peng, D. Y., and Robinson, D. B.: "A New Two-Constant Equation-of-State," *Ind. and Eng. Chem. Fund.* (1976) **15,** 59–64.

Perry, G. E.: "Weeks Island S Sand Reservoir B Gravity Stable Miscible CO_2 Displacement, Iberia Parish, Louisiana, Third Annual Report," DOE/METC-5232-4, June 1980.

Peterson, A. V.: "Optimal Recovery Experiments with N_2 and CO_2," *Pet. Eng. Intl.* (Nov. 1978) 40–50.

Rathmell, J. J., Stalkup, F. I., and Hassinger, R. C.: "A Laboratory Investigation of Miscible Displacement by Carbon Dioxide," paper SPE 3483 presented at SPE 4th Annual Fall Meeting, New Orleans, October 3–6, 1971.

Reamer, H. H., Olds, R. H., Sage, B. H., and Lacey, W. N.: "Phase Equilibria in Hydrocarbon Systems," *J. Ind. Eng. Chem.* (Jan. 1944).

Redlich, O., and Kwong, J. N. S.: *Chem. Rev.* (1949) **44,** 233.

Reid, R. C., Prausnitz, J. M., and Sherwood, T. K.: *The Properties of Gases and Liquids,* third edition, McGraw-Hill Book Co., New York (1977).

Reid, T. B., and Robinson, H. J.: "Lick Creek Meakin Sand Unit Immiscible CO_2—Waterflood Product," *J. Pet. Tech.* (Sept. 1981) 1723–1729.

Robinson, D. B., and Mehta, B. R.: "Hydrates in the Propane–Carbon Dioxide–Water System," *J. Can. Pet. Tech.* (Jan.–March 1971) **33.**

Rosman, A., and Zana, E.: "Experimental Studies of Low IFT Displacement by CO_2 Injection," paper SPE 6723 presented at the SPE 52nd Annual Technical Conference, Denver, October 9–12, 1977.

Ross, G. D., Todd, A. C., Tweddie, J. A., and Will, A. G. S.: "The Dissolution Effects of CO_2-Brine Systems on the Permeability of U.K. and North Sea Calcareous Sandstones," paper SPE/DOE paper 10685 presented at the 3rd Joint SPE/DOE Symposium on Enhanced Oil Recovery, Tulsa, April 4–7, 1982.

Rutherford, W. M.: "Miscibility Relationships in the Displacement of Oil by Light Hydrocarbons," *Soc. Pet. Eng. J.* (Dec. 1962) 340–346.

Shelton, J. L., and Schneider, F. N.: "The Effects of Water Injection on Miscible Flooding Methods Using Hydrocarbons and Carbon Dioxide," *Soc. Pet. Eng. J.* (June 1975) 217–226.

Shelton, J. L., and Yarborough, L.: "Multiple Phase Behavior in Porous Media During CO_2 or Rich-Gas Flooding," *J. Pet. Tech.* (Sept. 1977) 1171–1178.

Sigmund, P. M., Aziz, K., Lee, J. I., Ngiem, L. X., and Mehra, R.: "Laboratory CO_2 Floods and Their Computer Simulation, Developments in Reservoir Engineering," *Proc.,* 10th World Pet. Cong., Bucharest (1979) **3,** 243–250.

Simon, R., and Graue, D. J.: "Generalized Correlations for Prediction Solubility, Swelling and Viscosity Behavior of CO_2-Crude Systems," *J. Pet. Tech.* (Jan. 1965) 102–106.

Simon, R., Rosman, A., and Zana, E.: "Phase-Behavior Properties of CO_2-Reservoir Oil Systems," *Soc. Pet. Eng. J.* (Feb. 1978) 20–26.

Soave, G., "Equilibrium Constants from a Modified Redlich-Kwong Equation of State," *Chem. Eng. Sci.* (1972) **27**, 1197–1203.

Spivak, A., Perryman, T. L., and Norris, R. A.: "A Compositional Simulation Study of the SACROC Unit CO_2 Project," prepared for the 9th World Petroleum Congress, Tokyo, May 11–16, 1975.

Stalkup, F. I., Jr.: "Carbon Dioxide Miscible Flooding: Past, Present and Outlook for the Future," *J. Pet. Tech.* (Aug. 1978) 1102–1109.

Stalkup, F. I., Jr.: *Miscible Displacement,* Monograph Series, SPE, Dallas (1983) **8**, 137–158.

Standing, M. B.: "A Pressure-Volume-Temperature Correlation for Mixtures of California Oils and Gases," *Drill. and Prod. Prac., API* (1947) 275–286.

Stone, H. L.: "Estimation of Three-Phase Relative Permeability and Residual Oil Data," *J. Can. Pet. Tech.* (Oct.–Dec. 1973) 53–61.

Tiffin, D. L., and Yellig, W. F.: "Effects of Mobile Water on Multiple Contact Miscible Gas Displacements," paper SPE/DOE paper 10687 presented at the 3rd Joint SPE/DOE Symposium on Enhanced Oil Recovery, Tulsa, April 4–7, 1982.

Tumasyan, A. B., Panteleev, V. G., and Meinster, G. P.: "Effect of Carbon Dioxide on the Physical Properties of Petroleum and Water," *Nefteprom. Delo* (1969) 20.

Turek, E. A., Metcalfe, R. S., Yarborough, L., and Robinson, R. L.: "Phase Equilibria in Carbon Dioxide–Multicomponent Hydrocarbon Systems: Experimental Data and an Improved Prediction Technique," paper SPE 9231 presented at the 55th Annual Fall Technical Conference and Exhibition, Dallas, September 21–24, 1980.

Unruh, C. H., and Katz, D. L.: "Gas Hydrates of Carbon Dioxide-Methane Mixtures," *Trans., AIME* (1949) 83–86.

Vazquez, M., and Beggs, H. D.: "Correlations for Fluid Physical Property Prediction," *J. Pet. Tech.* (June 1980) 968–970.

Vukalovich, M. P., and Altunin, V. V.: *Thermophysical Properties of Carbon Dioxide,* Collet's Ltd., London (1968) 243–263, 351.

Wang, G. C., and Locke, C. D.: "A Laboratory Study of the Effects of CO_2 Injection Sequence on Tertiary Oil Recovery," *Soc. Pet. Eng. J.* (Aug. 1980) 278–280.

Warner, H. R.: "An Evaluation of Miscible CO_2 Flooding in Water-Flooded Sandstone Reservoirs," *J. Pet. Tech.* (Oct. 1977) 1339–1347.

Watkins, R. W.: "A Technique for the Laboratory Measurement of Carbon Dioxide Unit Displacement Efficiency in Reservoir Rock," paper SPE 7474 presented at the SPE 53rd Annual Technical Conference, Houston, October 1–4, 1978.

Watson, K. M., Nelson, E. F., and Murphy, G. B.: "Characterization of Petroleum Fractions," *Ind. and Eng. Chem.* (1935) **27**, 1460.

Whorton, L. P., Brownscombe, E. R., and Dyes, A. B.: "A Method for Producing Oil by Means of Carbon Dioxide," U.S. Patent No. 2,623,596 (1952).

Williams, C. A., Zana, E. N., and Humphrys, G. E.: "Use of the Peng-Robinson Equation of State to Predict Hydrocarbon Phase Behavior and Miscibility for Fluid Displacement," paper SPE 8817 presented at the 1st Joint SPE/DOE Symposium on Enhanced Oil Recovery, Tulsa, April 20–23, 1980.

Yarborough, L., and Smith, L. R.: "Solvent and Driving Gas Compositions for Miscible Slug Displacement," *Soc. Pet. Eng. J.* (Sept. 1970) 298–310.

Yellig, W. F.: "Carbon Dioxide Displacement of a West Texas Reservoir Oil," paper SPE/DOE 9785 presented at the 2nd Joint SPE/DOE Symposium on Enhanced Oil Recovery, Tulsa, April 5–8, 1981.

Yellig, W. F., and Metcalfe, R. S.: "Determination and Prediction of the CO_2 Minimum Miscibility Pressure," *J. Pet. Tech.* (Jan. 1980) 160–168.

Zudkevitch, D., and Jaffe, J.: "Correlation and Prediction of Vapor-Liquid Equilibria with the Redlich-Kwong Equation of State," *AIChE J.* (1970) **16**, 112.

4 RESERVOIR SIMULATION METHODS

The complexity of displacement mechanisms involved in the CO_2 injection process precludes any accurate prediction of reservoir performance by empirical means. In the past, adequate analysis of field response relied solely on information obtained from physical model studies. However, proper scaling of miscible displacements in laboratory models is difficult if not impossible to obtain. Consequently, considerable effort has been devoted to developing numerical models whose results accurately parallel the physical processes active in a reservoir under CO_2 injection.

During the 1970s, the art of reservoir modeling improved dramatically. This has been a direct result of advances made in computer hardware and software. It is now possible to write sophisticated numerical models that simulate some of the most complex phenomena that take place in different phases of petroleum recovery.

This chapter provides a general description of the three reservoir simulation methods used in connection with CO_2 flooding:

1. immiscible black oil modeling,
2. compositional modeling, and
3. miscible black oil modeling.

It is important to note that none of these simulators models all the factors that prevail in a given CO_2 displacement. However, the chapter discusses their applications as well as their strengths and weaknesses.

4.1   **THE NATURE OF RESERVOIR SIMULATION.** Almost all reservoir simulation applications are based upon the conservation of mass or heat in a system that exists in a fixed space and time domain. This system is separated from the surroundings by some physical boundaries. In reservoir simulation, these external boundaries are most frequently no flow boundaries. Fluid injection and production take place at wells that compose the internal boundaries of the system. These wells can be represented by point sinks and sources, and they will appear in the mathematical formulation as forcing functions. As a concluding remark on the boundary conditions, it is obvious that anything that enters or leaves the system must cross the boundaries (external and/or internal) of the system.

From figure 4.1 it is clear that certain rock and fluid properties are functions of time, space (x, y, z coordinates), and time- and/or space-dependent (pressure, saturation terms) variables. Therefore, at some initial time the description of the system by some set of conditions will be necessary. This set of conditions will compose the initial conditions of the problem.

After defining the initial and boundary conditions of the problem, it is necessary to state a correct description of the processes that are occurring in the system. Recognition of certain physical principles that control the process is of ultimate importance to the construction of a mathematical model. These physical principles will include conservations of mass, energy, and momentum; EOSs; and equations that describe the flow in porous media.

Engineers use two distinct approaches in the development of mathematical models in petroleum reservoir engineering:

1. a multiphase or single-phase flow system where mass balances are performed over each individual hydrocarbon component *(compositional model)* and
2. a multiphase or single-phase flow model where the hydrocarbon system is approximated by two components, a nonvolatile component (oil) and a volatile (gas) component *(black oil model)*.

The second approach, black oil modeling, is a subset of the more complex compositional approach. Thus, we will start with the formulation of compositional simulation and, later, will collapse this formulation to the black oil model case.

4.2   **EQUATION OF CONTINUITY.** Almost all reservoir simulation applications are based upon the conservation of mass or heat. Consider a small elemental volume around which a material balance can be expressed simply as

Mass in $-$ mass out $+$ net change $= 0$.

It should be noted that the elemental volume shown in figure 4.2A is chosen as a rectangular prism simply because the development will be carried out in a rectangular coordinate system. It is obvious that extension of this development to cylindrical and spherical coordinate systems is straightforward and can be done in a similar manner provided one starts with a cylindrical or spherical elemental volume (fig. 4.2B).

Consider a small element of reservoir volume, with sides Δx, Δy, and Δz and porosity ϕ, and consider flow only in the x direction for a moment. The rate of mass flowing in through the face at x is $(\rho v_x)|_x \Delta y \Delta z$, and similar terms can be written for the

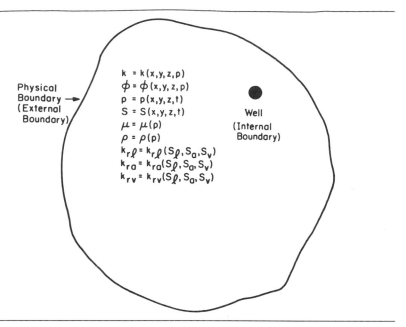

Figure 4.1 Physical system.

other two pairs of faces. Since the rate of mass accumulation in the volume element is $\frac{\delta}{\delta t}(\rho\phi\Delta x\Delta y\Delta z)$, the mass balance on the volume element is

$$[(\rho v_x)|_x - (\rho v_x)|_{x + \Delta x}]\Delta y\Delta z + [\text{similar for } y \text{ and } z]$$

$$= \frac{\delta}{\delta t}(\rho\phi)\ \Delta x\Delta y\Delta z. \qquad (4.1)$$

Dividing through this equation by $\Delta x\Delta y\Delta z$ and taking the limit as these dimensions approach zero gives

$$-\frac{\delta}{\delta x}(\rho v_x) - \frac{\delta}{\delta y}(\rho v_y) - \frac{\delta}{\delta z}(\rho v_z) = \frac{\delta}{\delta t}(\rho\phi). \qquad (4.2)$$

This is the equation of continuity that describes the rate of change of density at a fixed point due to the changes in the mass velocity vector $\rho\vec{v}$. In vector notation, equation (4.2) becomes

$$-\nabla \cdot (\rho\vec{v}) = \frac{\delta}{\delta t}(\rho\phi). \qquad (4.3)$$

Noting that the derivative operator has the units of reciprocal length, the units of $\nabla\ \cdot$

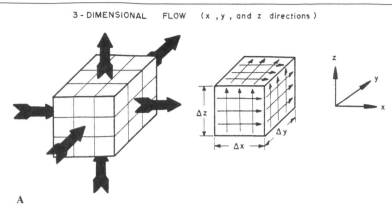

A

Figure 4.2A Rectangular elemental volume and material balance.

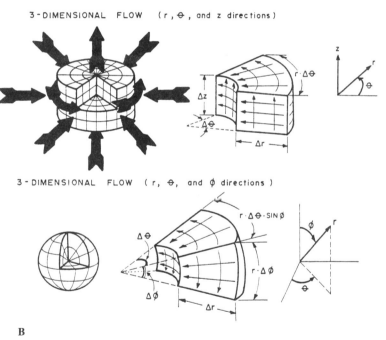

B

Figure 4.2B Elemental volumes of cylindrical and spherical flow geometries.

$(\rho\vec{v})$ are $(cm^{-1})(gm/cc)(cm/sec)$ or $(gm/sec)/cc$. Thus, the left side of equation (4.3) is simply the net rate of mass efflux per unit volume. Therefore, the equation says that the rate of increase in density in a small volume element is equal to the rate of mass influx into the element divided by the volume.

This equation contains all the physics required to describe reservoir fluid flow, provided that the density and velocity are defined properly in various situations. In the following paragraphs, the notation in the flow equation will be expanded to emphasize details of the flow, but the meaning does not change.

To write the equation of continuity more specifically for hydrocarbon reservoir fluid flow, three further considerations should be included—namely, that the flow can be multiphase, multimechanism, and multicomponent. *Multiphase* means that the flow involves mechanically distinct fluids (i.e., immiscible fluids), consisting of any combination of the oil phase, the gas phase, and the water phase. Each phase has its own distinct velocity and density, so the $\rho\vec{v}$ and $\rho\phi$ terms must be expanded. Let us consider for simplicity a two-phase, gas-oil system and ignore the reservoir water. Keeping in mind that $\rho\vec{v}$ was the rate of total mass influx into the volume element, for two-phase flow this becomes

$$\rho\vec{v} = \rho_o\vec{v}_o + \rho_g\vec{v}_g, \tag{4.4}$$

where the subscripts o and g refer to oil and gas respectively. Likewise, the density term becomes

$$\phi\rho = \phi S_o\rho_o + \phi S_g\rho_g; \tag{4.5}$$

that is, the oil density applies to that part of the volume containing the oil and similarly for the gas. Thus, the multiphase equation of continuity is

$$-\nabla \cdot (\rho_o\vec{v}_o + \rho_g\vec{v}_g) = \frac{\delta}{\delta t}(\phi S_o\rho_o + \phi S_g\rho_g), \tag{4.6}$$

with the auxiliary equation $S_g + S_o = 1$.

Multicomponent means that the flow involves more than one separately identifiable species, or component. In the absence of chemical reaction between species, an equation of continuity must hold separately for each component. Denoting the mass fraction of component i in the oil phase by ω_{oi} and similarly ω_{gi} for the component i in the gas phase, the equation of continuity for this component is

$$-\nabla \cdot (\rho_o\omega_{oi}\vec{v}_{oi} + \rho_g\omega_{gi}\vec{v}_{gi}) = \frac{\delta}{\delta t}(\phi S_o\rho_o\omega_{oi} + \phi S_g\rho_g\omega_{gi}), \tag{4.7}$$

where $\vec{v}_{oi}$ is the velocity of component i in the oil phase, similarly for $\vec{v}_{gi}$, and of course for n components there are n such equations. With the introduction of the mass fraction into these equations, the constraints $\sum_i \omega_{oi} = 1$ and $\sum_i \omega_{gi} = 1$ apply. Some other relationship between the ω_{oi} and ω_{gi} must be specified in order to solve these equations.

Finally, *multimechanism* means that the flow is occurring due to more than one mass transport mechanism. Under most circumstances, the mass transport due to convection is dominant, but in some cases the mass flow due to dispersion is significant. This is accounted for in equation (4.7) by recognizing that the velocity of component i in the oil phase, $\vec{v}_{oi}$, is the sum of velocities due to different mechanisms; for example,

$$\vec{v}_{oi} = \vec{v}_o^c + \vec{v}_{oi}^d, \tag{4.8}$$

where the superscripts c and d denote convection and dispersion respectively.

Note that the convective velocity was written without the subscript i. This is because convective velocity is associated with movement of the entire phase, by definition. In fact, we can write

$$\vec{v}_o^c = \sum_{i=1}^{n} \omega_{oi}\, \vec{v}_{oi}, \tag{4.9}$$

which says that the convective velocity is the mass-weighted average of the component velocities in that phase. By writing the velocity for each component as a sum of velocities, the continuity equation for component i becomes

$$-\nabla \cdot (\rho_o \omega_{oi} \vec{v}_o^c + \rho_g \omega_{gi} \vec{v}_g^c)$$

$$-\nabla \cdot (\rho_o \omega_{oi} \vec{v}_{oi}^d + \rho_g \omega_{gi} \vec{v}_{gi}^d) \tag{4.10}$$

$$= \frac{\delta}{\delta t}(\phi S_o \rho_o \omega_{oi} + \phi S_g \rho_g \omega_{gi}).$$

Note that the distinction of separate mass transport mechanisms, be it convection and/or dispersion, had no effect on the accumulation term.

Although equation (4.10) looks considerably more complicated than the original equation of continuity, it is still only the equation of continuity, and its meaning is not any more complicated than that of the original. However, the notation has been intended to illustrate that the flow is both convective and dispersive in each of the oil and gas phases for a given component i.

By making different simplifying assumptions, equation (4.10) can be reduced to the more familiar equations of miscible displacement models, compositional models, and black oil models.

4.3   MODELING THE CARBON DIOXIDE DISPLACEMENT PROCESS

4.3.1   Black Oil Models. The major assumptions in black oil models are

multiphase flow (gas, oil, and water);
up to three components called gas, oil, and water; and
negligible dispersion.

Further, it is usual to assume that the oil component can exist only in the oil phase, that the gas component can exist in both the gas and oil phases but not in the water phase, and that the water component can exist only in the water phase. Ignoring the water phase, these latter assumptions imply that the mass fractions of gas in the oil and gas phases (1 = gas component, 2 = oil component) are given by

$$\omega_{o1} = \frac{m_g}{m_o + m_g}; \qquad \omega_{g1} = 1, \tag{4.11}$$

and the mass fractions of oil in the oil and gas phases are

$$\omega_{o2} = \frac{m_o}{m_o + m_g}; \qquad \omega_{g2} = 0, \tag{4.12}$$

where m_g is the mass of the gas component in the oil phase and m_o is the mass of the oil component in the oil phase.

Using the definitions of oil and gas formation volume factors,

$$B_o = \frac{V_o \rho_{oSTC}}{m_o} \qquad V_o = \frac{m_o + m_g}{\rho_o} \tag{4.13}$$

and

$$B_g = \frac{\rho_{gSTC}}{\rho_g}, \tag{4.14}$$

and of solution gas-oil ratio,

$$R_s = \frac{V_{gSTC}}{V_{oSTC}} = \frac{m_g}{m_o} \cdot \frac{\rho_{oSTC}}{\rho_{gSTC}}, \tag{4.15}$$

then

$$\rho_o \omega_{o1} = \frac{R_s}{B_o} \rho_{gSTC}, \tag{4.16}$$

and

$$\rho_o \omega_{o2} = \frac{1}{B_o} \rho_{gSTC}. \tag{4.17}$$

Last, neglecting dispersion, equation (4.10) reduces to the gas and oil component equations

$$-\nabla \cdot \left(\frac{R_s}{B_o} \vec{v}_o^c + \frac{1}{B_g} \vec{v}_g^c \right) = \frac{\delta}{\delta t} \left(\phi S_o \frac{R_s}{B_o} + \phi S_g \frac{1}{B_g} \right) \tag{4.18}$$

and

$$-\nabla \cdot \left(\frac{1}{B_o}\vec{v}_o^c\right) = \frac{\delta}{\delta t}\left(\phi S_o \frac{1}{B_o}\right). \tag{4.19}$$

Along with the auxiliary two-phase equations, these form a closed set. It is important to note that while typical black oil models exclude natural gas solubility in the water phase, this fact cannot be neglected in CO_2 displacements. It is quite a simple matter to include R_{sw} in the derivation of equations (4.11) through (4.19).

4.3.2   Compositional Models. The usual assumptions involved in compositional models are

multiphase flow,
two or more components, and
negligible dispersion.

The assumption of negligible dispersion reduces the general component equations to

$$-\nabla \cdot (\rho_o \omega_{oi}\vec{v}_o^c + \rho_g \omega_{gi}\vec{v}_g^c) = \frac{\delta}{\delta t}(\phi S_o \rho_o \omega_{oi} + \phi S_g \rho_g \omega_{gi}),$$

$i = 1$ to n components. $\tag{4.20}$

To form a complete set of equations, the mass fractions ω_{gi} and ω_{oi}, are assumed to be related by the phase equilibrium coefficients, the so-called K-values; that is,

$$\frac{\omega_{gi}}{\omega_{oi}} = K_i = f(\rho,\omega_i), \qquad i = 1 \text{ to } N. \tag{4.21}$$

The K-values may be known functions of pressure and composition, or they can be the solution of a nonlinear set of equations derived from some EOS. And, of course, the convective velocities are usually assumed to be the Darcy phase velocities,

$$\vec{v}_o^c = -\frac{k\,k_{ro}}{\mu}(\nabla P_o - \gamma\nabla z), \tag{4.22a}$$

$$\vec{v}_g^c = -\frac{k\,k_{rg}}{\mu}(\nabla P_g - \gamma\nabla z), \tag{4.22b}$$

and

$$P_{cgo} = P_g - P_o. \tag{4.23}$$

With various constraining equations, this discussion forms a complete set of $2N + 4$ equations with $2N + 4$ unknowns P_o, P_g, S_o, S_g, and ω_{oi}, ω_{gi}, $i = 1$ to N.

4.3.3   **Miscible Displacement.** The usual assumptions involved in miscible displacement are

single-phase flow;
two components, called oil and solvent; and
significant dispersion.

In the following, the subscripts 1 and 2 are used for the oil and solvent components to avoid confusion with phase subscripts. The general equation for the oil component is

$$-\nabla \cdot (\rho\omega_1 \vec{v}^c + \rho\omega_1 \vec{v}_1^d) = \frac{\delta}{\delta t}(\phi\rho\omega_1) \tag{4.24}$$

and similarly for solvent. The dispersive term is usually replaced by Fick's law; namely,

$$\rho\omega_1 \vec{v}_1^d = -K\nabla(\rho\omega_1), \tag{4.25}$$

which states that the mass rate of flow of oil due to dispersion is proportional to the mass concentration gradient, the proportionality being the dispersion coefficient tensor, K. Assuming that the density of oil and solvent (D_1 and D_2) are both constant, equation (4.25) can be expressed in terms of volumetric concentration of solvent defined by

$$C = \frac{\rho\omega_2}{D_2} = 1 - \frac{\rho\omega_1}{D_1}. \tag{4.26}$$

The resulting equations with ω_1 and ω_2 replaced by C are

$$-\nabla \cdot (1 - C)\vec{v}^c - K\nabla(1 - C) = \frac{\delta}{\delta t}\phi(1 - C) \tag{4.27}$$

and

$$-\nabla \cdot (C\vec{v}^c - K\nabla C) = \frac{\delta}{\delta t}\phi C.$$

It is usual to add the equations, with the result

$$\nabla \cdot \vec{v}^c = \frac{\delta}{\delta t}(\phi). \tag{4.28}$$

Expressing $\vec{v}^c$ in terms of Darcy's law, these last two equations then form a complete set of unknowns: p and C.

4.4   **APPLICATIONS, STRENGTHS, AND WEAKNESSES OF SIMULATION MODELS.** The basic difference between the three different types of simulation models—compositional, black oil immiscible, and black oil miscible—lies in their treatment of fluid and phase properties. The assumptions involved in compositional models are multiphase flow, two or more components, and negligible dispersion. The assumptions in black oil immiscible models are multiphase flow (gas, oil, and water); up to three components called gas, oil, and water; and negligible dispersion. Finally, a miscible model assumption involves single phase, usually two first-contact miscible components, and significant dispersion.

4.4.1   **Black Oil Immiscible Model.** Traditional black oil simulation is based on the solution of Muskat's (1937) original equation of continuity (Blackwell et al. 1959). West et al. (1954) presented the first numerical solution techniques of the gas and oil equations. Douglas et al. (1959) then coupled the flow equations (IMPES) for sequential solution of pressures and then saturations.

For CO_2 injection, traditional black oil simulators may be adequate in predicting performance of *low pressure, immiscible displacements* only. The pressure-temperature region of applicability of immiscible CO_2 injection is shown approximately by Region I on figure 4.3.

Black oil models are capable of simulating systems where water, oil, and CO_2 are present in all proportions. However, if a fourth fluid, natural gas, is present in significant quantities, this method may fail due to its inability to account for mixing of gases. Usually, phase transfers are accounted for between the gas and water and the gas and oil phases by solution gas-oil-water ratio functions. There is no vaporization of liquid components. Fluid properties in this type of simulator are obtained from the oil formation volume factor, water formation volume factor, gas formation volume factor, phase viscosities, solution gas-oil ratio, and solution gas-water ratio. These are all specified functions of phase pressure only.

Four papers have addressed the application of black oil simulation in low pressure, immiscible displacements. Stright et al. (1977), Klins et al. (1982), Reid and Robinson (1981) and Patton et al. (1982) have investigated the injection of CO_2 in moderately viscous reservoirs as either a cyclic stimulation or a drive process.

4.4.2   **Black Oil Miscible Model.** Section 4.3.3 discussed the general formulation of equations to describe the miscible displacement of oil by a solvent, where the oil and solvent are miscible in any mixture proportion. The major stumbling block of these convection/diffusion-type models seems to be in the solution scheme where numerical dispersion at the flood front appears to be much greater than the actual diffusional mixing of oil and solvent.

Peaceman and Rachford (1962) presented the first numerical solution scheme of the miscible displacement (convection/diffusion model) process. They found, however, to

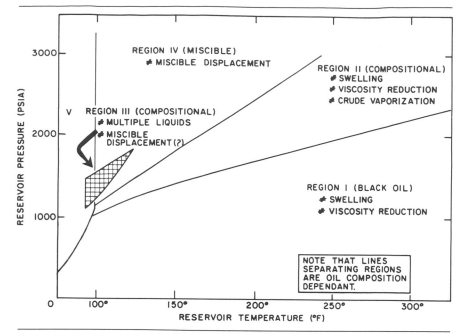

Figure 4.3 The effect of reservoir temperature and pressure on CO₂ injection displacement mechanisms and the applicable simulation techniques.

overcome numerical dispersion of the flood front in their one-dimensional flow model, a minimum of 48 grid blocks was necessary. Even with such a large number of blocks, the length of the solvent/oil mixing zone in the region immediately behind the flood front was longer (smeared by numerical dispersion) than calculated by an exact solution technique.

A number of other authors including Gardner et al., Chaudhari (1971), and Larson (1982) have presented alternatives for controlling numerical dispersion in the convection/diffusion, miscible model. Most have met with little success.

A parallel track was taken to design a simulator for the first-contact, always miscible process by altering rock and fluid property relationships in a traditional black oil immiscible simulator (Lantz 1970). By adjusting the capillary pressure and relative permeability relationships in a two-phase (oil and gas) black oil model, Lantz found first-contact miscible displacement and dispersion could be simulated. This approach, however, also failed to predict accurately the solvent/oil mixing effect at the flood front when there was a significant viscosity difference between the solvent and oil. Such an effect is present in the CO₂ miscible displacement case.

Viscosity differences lead to fingering of the less viscous CO₂ through the oil. This effect, shown earlier in figure 1.10, makes it very difficult to predict finger location accurately when grid blocks are considerably larger than the solvent fingers they are attempting to model. As a finger of solvent moves into a grid block, it mixes instantaneously with the oil in the block, creating an oil/solvent mixture that moves more slowly out of the grid block than low viscosity, pure solvent (in this case, CO_2) would. Rather than predicting jagged fingers of CO_2 protruding into the oil zone, the velocity of these fingers is slowed, making a smeared flood front. Previous models that failed to account for CO_2 fingering led to delayed CO_2 breakthrough, increased sweep efficiency, and overly optimistic predictions of oil recovery.

Koval (1963) developed an analytical solution to predicting the performance of these unstable, miscible displacements. Todd and Longstaff (1972) extended this work to develop a modified black oil simulator for first-contact miscible displacement where viscous fingering and areal sweep dominate. Like Lantz, they modified the gas-oil relative permeability curves for miscible flow and presented a method of calculating effective viscosities of oil and gas phases when mixing occurs. These effective viscosity relationships (to be discussed in more detail later) allowed the modeler to adjust the velocity of the solvent and oil phases by altering one mixing constant.

Unlike the true convection/diffusion models, the modified black oil miscible models do not attempt to calculate true physical mixing of solvent and oil. Instead, they attempt to predict the results of mixing and miscible displacement by adjustment of an empirical mixing parameter.

These miscible-type black oil models (Koval or Todd-Longstaff) involve a simplification of phase properties. The fluids are always assumed to be miscible on contact as was described earlier as first-contact or very rapid multiple-contact miscibility (high in Region IV of fig. 4.3).

Also, the density and viscosity of the mixture depend primarily on solvent concentration, and the formulation assumes no volume change on mixing. Explicit treatment of dispersion makes miscible models useful in situations where recovery is controlled by sweep efficiency and the interaction of viscous fingering. Dispersion is considered a critical mechanism of displacement, while phase behavior (mass transfer) is not.

If you have a process that is completely miscible on first contact or in which the distance required to obtain miscibility is short with respect to well spacing, the major problem in the application of this scheme lies in the selection of the appropriate dispersion mixing parameter; the Todd-Longstaff ω or Koval's K. Other data required parallel those used in black oil simulation. Permeability, relative permeability, porosity, and capillary pressure as well as solely pressure-dependent variables—viscosity, formation volume factor, solubilities, and densities—are needed.

4.4.3   **Compositional Model.** Compositional simulation is the natural choice for CO_2 flooding applications where complex phase behavior and compositionally dependent phase properties such as density, interfacial tension, and viscosity are more critical to predicting displacement efficiency than viscous fingering and areal sweep effects; that is, while modeling phase behavior effects, dispersive mass transfer (viscous fingering) as a rule is neglected in traditional composition models.

When there is considerable mass transfer between the flowing phases, use of a multicomponent compositional model is dictated. This type of phase transfer is typical of the vaporization or condensation processes that take place in Regions II, III and the lower section of Region IV on figure 4.3. Multiple-contact developed miscibility processes like these cannot be formulated accurately on modified black oil simulators.

However, compositional simulation is most complex since it focuses on the transport of individual components rather than phases through a porous medium. Solution of individual block compositions relies on correct partitioning of individual components into the oleic and vapor phases. Nolen (1973) and Kazemi et al. (1978) have introduced compositional simulators that distribute the components among phases using convergence pressure correlations for equilibrium K-values. Fussell and Fussell (1979), Coats (1980), and Nghiem et al. (1981) have presented models that rely on an EOS to separate components. Regardless of the technique (convergence pressure or EOS), compositional simulation introduces a significant increase in the number of variables over black oil immiscible and miscible models. This may introduce severe computational time requirements if the number of components used to match phase behavior is large.

Time and storage constraints are further tested if a large number of grid blocks are employed to reduce viscous fingering and dispersion effects. Also, with a multiple-contact miscible process, miscibility should occur at some unknown distance between injection and production wells. Even with a large number of grid blocks, one may not be able to match accurately the point in the reservoir where miscibility is attained. However, regardless of these large time and storage requirements, there is no tractable replacement to predict performance of the multiple-contact miscible CO_2 process. With the advent of improved computer hardware (vector computers, etc.), time and storage restrictions may be reduced significantly.

If the distance to develop miscibility is short, a black oil model that assumes instantaneous miscibility between oil and CO_2 would be a wise choice. It requires fewer data, computer time, and storage to employ than the more complex compositional simulator. However, it should predict higher than actual oil recoveries by assuming first-contact miscibility.

Compositional simulators, while requiring more time and data than their black oil counterpart, are irreplaceable in predicting oil recovery for a CO_2-multiple-contact miscible process where miscibility occurs some distance from the injection well. It is also the only choice in low temperature applications ($T_{res} < 122°F$) if multiple liquids are formed.

4.4.4   **Hybrid Miscible/Compositional Model.** Chase and Todd (1982) and Ko (1982) have introduced compositional simulators that mesh the best features of the black oil miscible and compositional models. They do not attempt to partition components in as exacting a fashion as a traditional compositional model, and the segregation of fluids is controlled through the mixing parameter approach.

Other model features can include

neglecting component transfer when block pressure exceeds a preset miscibility pressure,
oil blocking by a mobile water phase,
an immobile residual oil saturation,
solubility of CO_2 in the water phase, and
asphaltene dropout.

While not determining implicitly the phase behavior in the multiple-contact region, this model does introduce a quasi-compositional approach that may require 10 to 100 times less storage and run time than the fully compositional approach. However, such an approach requires a substantial amount of data over traditional compositional modeling. Data particularly useful for this sort of simulation are minimum miscibility pressure (slim tube), solubility of CO_2 in oil and water, K-values for other light components, gas super compressibilities, inaccessible and remaining miscible oil saturation, as well as the proper choice of mixing parameter.

In review of the previous discussion of model applications, it is important to note that no model exists that represents all the salient features of CO_2 displacements. Immiscible black oil models are limited to a small pressure-temperature range of applications while compositional simulators mirror the phase transfer process and neglect the effect of viscous fingering and numerical dispersion. Miscible models, on the other hand, ignore phase behavior for the sake of predicting fingering effects. Hybrid models lie somewhere in the middle, with some phase transfer, dispersive mixing, and large data requirements. The type of CO_2 process planned will determine the type of model to be used. Table 4.1 presents the applications of these models by several authors for field planning.

4.5   COMPARISON OF KOVAL AND TODD-LONGSTAFF MISCIBLE MODELS

4.5.1   Mixing Parameters for the Black Oil Miscible Model. Because of the complexity of the solution of a set of generalized compositional equations, there was strong motivation to develop first-contact miscible simulation models using the conventional black oil model premise. Early work by Koval (1963), as presented in chapter 2, pioneered the way of predicting miscible displacement performance long before the advent of numerical simulation. His method, analogous to the Buckley-Leverett technique, predicted oil recovery and solvent cut as a function of the pore volumes of solvent injected.

In 1972, Todd and Longstaff extended Koval's (as well as a number of other authors') work in developing a *black oil simulator* for predicting first-contact miscible flood performance (Chaudhari 1971; Gardner et al. 1964; Lantz 1970; Peaceman and Rachford 1962; Perrine and Gay 1966; Price and Donahue 1967). They introduced the mixing parameter, ω, to determine the average viscosity and density of the mixture in the dispersed zone between solvent and oil. This feature allowed the use of coarser grids in reservoir simulation without masking the effects of dispersion at the flood front. Their approach is discussed in relation to Koval's earlier work.

Koval's method accounts for fingering of solvent or bypassing of oil by utilizing an effective viscosity ratio, E, related to the actual viscosity ratio. His method gives a fractional flow of solvent (gas in this case) as

$$f_g = \frac{ES_g}{1 + S_g(E - 1)} \tag{4.29}$$

where

$$E = \left[0.78 + 0.22 \left(\frac{\mu_o}{\mu_g}\right)^{1/4}\right]^4,$$

$$S_g = \text{volume fraction of solvent.}$$

(4.30)

Independent work by Gardner and Ypma (1982) using core flood experiments and high-resolution two-dimensional simulation, also investigated the viscous fingering effects of unstable displacements of Wasson crude (no mobile water) by CO_2. Comparisons between their laboratory work and the oil recovery predicted from Koval's theory was quite reasonable.

Let us now derive a direct relationship between Koval's mixing parameter, K, and Todd-Longstaff's black oil model coefficient, ω. Note that water is ignored because it does not affect the relationship between K and ω.

In Koval's method, the interblock flows of total fluid and solvent (gas) are given by

$$q_T = \frac{kA}{L} \frac{1}{\mu} \Delta p$$

(4.31)

and

$$q_g = \frac{kA}{L} \frac{f_g}{\mu} \Delta p,$$

(4.32)

where μ is mixture viscosity and f_g is fractional flow of the solvent. Equating Koval's equation (4.29) with equations (4.31) and (4.32), the fractional flow of gas is defined as

$$\left[f_g = \frac{KS_g}{1 + S_g(K - 1)} = \frac{q_g}{q_T}\right]$$

(4.33)

By definition,

$$K = HE = H\left[0.78 + 0.22\left(\frac{\mu_o}{\mu_g}\right)^{1/4}\right]^4,$$

(4.34)

where μ_o and μ_g are the oil and solvent (gas) viscosities respectively, H is the heterogeneity factor (equal to one for homogeneous systems), and E is Koval's effective viscosity ratio.

The Todd-Longstaff method gives total and solvent interblock flows as

$$q_T = \frac{kA}{L}(\lambda_g + \lambda_o)\Delta p$$

(4.35)

and

$$q_g = \frac{kA}{L}\lambda_g \Delta p,$$

(4.36)

Table 4.1 Parametric and Field Test Modeling Studies

Author(s)	Model Type	Mode Geometry	Number of Phases	Grid Size	Comments
Reid and Robinson (1981)	Black oil	Three dimensional	3	—	
Stright et al. (1977)	Black oil	Two-dimensional cross section	3	7 × 10	
Klins and Farouq Ali (1982)	Black oil	Two-dimensional areal	3	9 × 9	
Patton et al. (1982a; 1982b)	Black oil	One-dimensional radial	3	20	
Pontious and Tham (1978)	Miscible	Two-dimensional areal	3	29 × 17	ω = 0.5
		Two-dimensional cross section	3	9 × 13	
Youngren and Charlson (1980)	Miscible	Two-dimensional areal	2	6 × 16	ω = 0.7
		Two-dimensional cross section	2	7 × 10	

Reference	Model	Dimensionality		Grid	Remarks
Bilhartz et al. (1978)	Miscible	Two-dimensional cross section	3	7×13	$\omega = 0.625$ to 0.75
Warner (1977)	Miscible	Two-dimensional cross section	3	25×5	$\omega = 0.8$
Desch et al. (1982)	Compositional	Three-dimensional	3	$16 \times 15 \times 4$	EOS K-values
Fayers et al. (1981)	Compositional	Two-dimensional cross section	3	6×12	EOS K-values
Graue and Zana (1981)	Compositional	One-dimensional	—	50	Convergence pressure K-values
Spivak et al. (1975)	Compositional	Two-dimensional cross section	3	20×10	Convergence pressure K-values
Chase and Todd (1982)		Three-dimensional		$10 \times 10 \times 10$	
Todd et al. (1982)	Hybrid	Two-dimensional cross section	3	$9 \times 3, 6$ $9 \times 9 \times 3$	$\omega = 0.625$
Ko (1982)	Hybrid	One dimensional	3	11	
		Two-dimensional cross section	3	$11 \times 2, 3, 5$	$\omega = 0.768$ to 0.930
		Three dimensional	3	$11 \times 2 \times 2$ $5 \times 5 \times 2$ $6 \times 6 \times 2$	

where

$$\lambda_g = S_g/(\mu_m^{\omega}\mu_g^{1-\omega}) \tag{4.37}$$

and

$$\lambda_o = (1 - S_g)/(\mu_m^{\omega}\mu_o^{1-\omega}), \tag{4.38}$$

and the mixed viscosity is defined as

$$\frac{1}{\mu_m^{1/4}} = \frac{S_g}{\mu_g^{1/4}} = \frac{1 - S_g}{\mu_o^{1/4}}. \tag{4.39}$$

The fractional flow of solvent in the miscible case is defined by Koval and Todd-Longstaff as follows:

$$f_g = \frac{KS_g}{1 + S_g(K - 1)} = \frac{\lambda_g}{\lambda_g + \lambda_o}, \tag{4.40}$$

where

$$\frac{\lambda_g}{\lambda_g + \lambda_o} = \frac{(S_g/\mu_m^{\omega}\mu_g^{1-\omega})}{S_g/\mu_m^{\omega}\mu_g^{1-\omega} + (1 - S_g)/\mu_m^{\omega}\mu_o^{1-\omega}} \tag{4.41}$$

and S_g and f_g are saturation and fractional flow of solvent.

The analogy between the Koval and Todd-Longstaff methods gives (proceed with $H = 1$)

$$K = \left(\frac{\mu_o}{\mu_g}\right)^{1-\omega} \tag{4.42}$$

so that either equation (4.34) or (4.42) may be used to evaluate K. Once K or ω is evaluated, either method, Koval or Todd-Longstaff, can be used to predict first-contact miscible (not multiple-contact miscible) performance.

The difficulty in using one of these two methods to predict first-contact miscible displacement phenomena lies in choosing a correct value of the mixing parameter K or ω. In using Koval's method, K can be directly determined using equation (4.34). However, one study by Paul et al. (1980) has shown that the use of these values to predict performance did not properly match laboratory data presented by Lacey et al. (1961). Figure 4.4 shows laboratory results that are matched when using a K-value approximately 1.33 times that calculated using equation (4.34).

A proper Koval's K could be estimated for this case using equation (4.42) and an omega value of 0.60. However, Todd and Longstaff found an omega value of 0.67 proper when forecasting the miscible flood performance of Blackwell et al. (1959) in the laboratory and a value of 0.33 for simulating full-scale displacements. A more recent

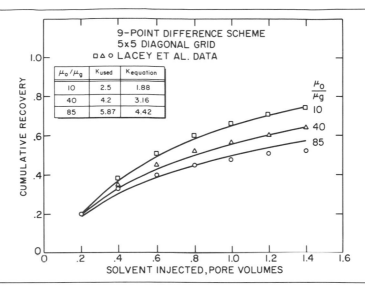

Figure 4.4 The effect of Koval's *K* mixing parameter on simulation results to match laboratory data (Lacey et al. 1961; Paul et al. 1980).

study by Warner (1977) estimated a correct omega value to be 0.80. A number of additional studies, shown in table 4.1, exhibit a wide variety of omega values used to match laboratory and pilot test data. This leads one to the initial conclusion that there is little agreement in choosing a correct mixing parameter value.

4.5.2   **Generalized Koval Method for Compositional Simulation.** As discussed earlier, traditional compositional modeling neglects dispersive mixing of CO_2 and oil. However, hybrid black oil/compositional models, discussed earlier, have been developed to include viscous fingering effects (*Compositional Reservoir Simulation* 1981). Consider the general *N*-component (hydrocarbon) case where original oil of given composition is displaced by an injected solvent of composition x_i^*. A constraint placed on component identities is that there must be at least one component that is present in the solvent but not in the original oil.

In the case of CO_2 injection, CO_2 may satisfy this constraint (if not present in the original oil). If a solvent of $C_1 - C_2 - C_3 - C_4$ were injected, then these components would likely all be present in the original oil. In that case, an extra component would have to be added such that component 1 = methane in solvent, component 2 = methane in original oil, component 3 = C_2, etc.

We express solvent fractional flow, following Koval, as

$$f = \frac{KF}{1 - S_{ot} + F(K - 1)}, \tag{4.43}$$

where F is volume or mole fraction of solvent in the grid block mixture,

$$S_{ot} = S_{ot}(S_w) \tag{4.44}$$

is a trapped oil saturation dependent upon water saturation, and

$$K = K_1 + K_2 F \tag{4.45}$$

is allowed to be either constant or a function of F. Note that f will equal 1 when $F = 1 - S_{ot}$ so that an oil saturation (or more precisely, mole fraction) of S_{ot} will remain undisplaced by solvent.

The problem now is to use equation (4.43) to express the effluent upstream or *flowing* composition, $\bar{x}_i$, to be used in the interblock flow expression for a compositional model. The upstream grid block composition is denoted by x_i, solvent composition by x_i^*. A simple expression for the appropriate flowing composition is given as

$$\bar{x}_i = \alpha x_i^* + (1 - \alpha)x_i. \tag{4.46}$$

The value of F is given by

$$F = x_i/x_i^*, \tag{4.47}$$

where it is assumed that component 1 is the component present in solvent but not in the original oil. The grid block oil is a mixture of solvent and oil of composition x_i^B,

$$x_i = Fx_i^* + (1 - F)x_i^B. \tag{4.48}$$

The flowing composition, by definition of f, is

$$\bar{x}_i = fx_i^* + (1 - f)x_i^B. \tag{4.49}$$

Substitution of x_i^B from equation (4.48) into (4.49) gives the desired expression

$$\bar{x}_i = \bar{f}x_i^* + (1 - \bar{f})x_i, \tag{4.50}$$

where

$$\bar{f} = \frac{f - F}{1 - F}, \tag{4.51}$$

and f is given by equation (4.43).

If K is less than 1, then equation (4.50) with (4.51) and (4.43) will result in numerical dispersion *control* or reduction, while an increase in K greater than one increases dispersion. Figure 4.5 summarizes this discussion.

The miscible displacement model described here will have some limitations. It should be recognized that this type of development does not include a rigorous treatment of thermodynamic and transport phenomena that describe the details of the fluid composi-

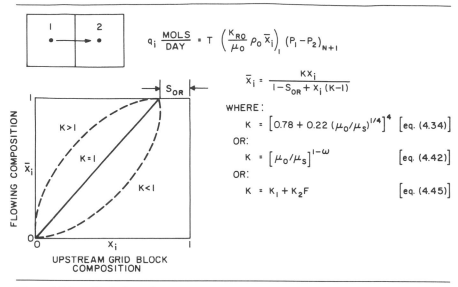

Figure 4.5 Interblock flow of components.

tion and flow characteristics. However, it allows for the inclusion of dispersion effects with compositional simulation of miscible flood performance in a very effective manner.

4.6 **MODELING PHASE BEHAVIOR IN INJECTION AND PRODUCTION WELLS.** When numerical reservoir simulators are employed for CO_2 project evaluation or design, a bottomhole boundary condition of constant pressure or constant flow is usually specified for injection and production wells. In fact, the bottomhole pressure changes in time with respect to flow rate, composition, and wellhead pressure. Usually, either injection wellhead or production separator pressure is controlled to a constant value, which means that an understanding of the wellbore phenomena is required for complete CO_2 project design.

Pressure drop, heat transfer, and phase behavior in the wellbore will control deliverability and injectivity. If the wellbore heats up (steam injection) or cools down (liquid CO_2 injection) with time, there is a long-term temperature transient that can also be important. As the reservoir's produced fluid composition changes, such as with CO_2/oil bank breakthrough into a producing well, the well's deliverability will change dramatically.

In the *injection* of CO_2, it is critical that bottomhole pressures are estimated with reasonable accuracy since all displacement performance is a function of pressure. Injecting CO_2 as a gas can be modeled by solution of the general momentum balance:

$$144\frac{dP}{dL} = \bar{\rho}\,\frac{g}{g_c}\sin\theta + \frac{\bar{f}\bar{\rho}v^2}{2g_c d} + \frac{\bar{\rho}v}{g_c}\left[\frac{dv}{dL}\right], \qquad (4.52)$$

where

P = pressure, psia,
L = pipe length, feet,
g = acceleration due to gravity, ft/sec^2,
g_c = gravitational constant, 32.17 ft/sec^2,
θ = inclination of pipe from vertical,
f = Moody friction factor,
$\bar{\rho}$ = average density, lbm/ft^3,
v = fluid velocity, ft/sec,
d = inside pipe diameter, ft.

Solution of this equation for single-phase flow involves (1) neglecting the velocity gradient, (2) estimating the average density as a constant for incompressible liquids or as a function of pressure (imperfect gas law) for gases, and (3) estimating the friction factor, f, with the equations presented in section 4.6.1.

4.6.1   Friction Factor for Single-Phase Flow. The Moody friction factor, f, is a function of the Reynolds number and the relative roughness of the pipe. The Reynolds number, N_{RE}, is a dimensionless number defined as

$$N_{RE} = \frac{d\,v\,\rho}{\mu}. \tag{4.53a}$$

For Reynolds numbers less than 2,000 flow is considered laminar and the friction factor can be estimated by

$$f = 64/N_{RE}. \tag{4.53b}$$

At Reynolds numbers greater than 2,000 the friction factor in smooth pipes can be estimated by

$$f = 0.0056 + 0.5N_{RE}^{-0.32}. \tag{4.53c}$$

In rough pipes, Colebrook's (1939) equation can be used, where

$$\frac{1}{\sqrt{f}} = 1.74 - 2\log_{10}\left(\frac{2\epsilon}{d}\right) + \frac{18.7}{N_{RE}\sqrt{f}}. \tag{4.53d}$$

By definition

μ = viscosity, lbm/ft-sec (cp $\times$ 0.000672),
ϵ = pipe roughness (distance from peaks to valleys in pipe wall irregularities), ft.

4.6.2   **Two-Phase Flow in Liquid Carbon Dioxide Injection.** For *two-phase vertical* injection (such may be the case when CO_2 is injected as a liquid), equation (4.52) has been reduced by Gould (1974) to

$$144\frac{dP}{dL} = \frac{\bar{\rho}\frac{g}{g_c} - \tau_f}{1 - \gamma_a},$$ (4.54a)

where

γ_a = acceleration gradient, dimensionless,
$\bar{\rho}$ = average density of mixture, lbm/ft³,
τ_f = friction loss gradient, lbf/ft² − ft.

The average density of the mixture may be computed as

$$\bar{\rho} = \rho_l H_L + \rho_g(1 - H_L),$$ (4.54b)

such that

ρ_L = liquid density, lbm/ft³,
ρ_g = gas density, lbm/ft³,
H_L = volume fraction liquid, holdup.

Frictional losses, τ_f, are a function of the flow regime present. Various flow patterns, as shown in figure 4.6, may all be present at different locations in a given wellbore at a given time. By calculating what Gould (1974) has described as the gas and liquid velocity influence number, the flow regime present at a given point can be estimated using figure 4.7. A number of authors have then presented correlations to calculate the frictional pressure losses dependent on the flow regime present. Table 4.2 gives a summary of available techniques.

If the injected fluid is steam or liquid CO_2, the average density used in equation (4.54a) becomes not only a function of pressure and temperature but also enthalpy. Solution of pressure drops in wellbores with phase changes requires the addition of a heat balance. The balance for liquid CO_2 injection is given by

$$Q + w\frac{dh_m}{dL} + w\frac{d}{dL}\left(\frac{V_m^2}{Zg_cJ_c} - \frac{wg}{g_cJ_c}\right) = 0,$$ (4.55)

where

Q = heat loss rate to surroundings, Btu/hr/ft,
w = mass flow rate, lbm/hr,
h_m = specific enthalpy of mixture, Btu/lbm,
V_m = mixture velocity, ft/sec,
J_c = mechanical equivalent of heat, 788 ft-lbt/Btu.

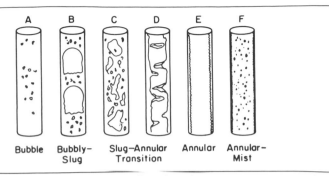

Figure 4.6 Flow patterns in concurrent vertical two-phase flow.

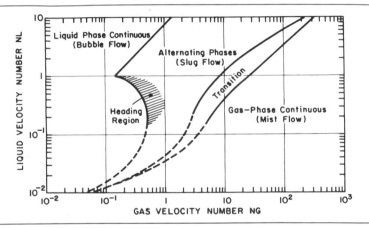

Figure 4.7 Extended flow regime map for vertical two-phase flow (Gould 1974).

Within the wellbore, heat transfer may occur by radiation, convection, and conduction. This is given by the following equation for radial heat flow:

$$Q = 2\pi r_{to} U(T - T_{cem}).$$ (4.56)

The overall heat transfer coefficient, U, may be calculated using the technique introduced by Willhite (1967).

Coupling of the heat and momentum balances then yields a nonlinear set of equations that must be solved simultaneously with iterative procedures typical of reservoir simulation. Such a model for estimating bottomhole properties of liquid or liquid-gas CO_2 injection has been presented by Cronshaw and Bolling (1982).

Using such a model, figure 4.8 shows that while injecting CO_2 at a surface pressure of 1,200 psia, bottomhole pressures may vary from 3,100 to 4,700 psia. The converse is also true. If a minimum pressure of 4,500 psia is needed for miscible displacement, tophole pressures could vary from 1,100 to 1,700 psia, depending on CO_2 injection properties.

Table 4.2 Multiphase Flow Correlations

Flow Regime	Model Author
Bubble	Griffith and Wallis (1961)
	Gould (1974)
	Beggs and Brill (1973)
	Aziz et al.
Slug	Chierici et al. (1980)
	Hagedorn and Brown (1965)
	Orkiszewski (1967)
	Beggs and Brill (1973)
	Aziz et al.
Transition	Duns and Ros (1963)
	Beggs and Brill (1973)
Annular mist	Duns and Ros (1963)
	Beggs and Brill (1973)

4.7 **APPENDIX: DERIVATION OF THE CONTINUITY EQUATION FOR COMPOSITIONAL SIMULATION.** Consider the general case where there are N components, each of which may exist in any or all of the gaseous, oleic, and aqueous phases. The conserved quantity entering the elemental volume is given as the summation of molar flow rate of component i in the aforementioned phases. Then, in this context, $N_{i\iota_x}$ (moles/day) will represent the molar flow rate of component i in the oleic phase in the x direction at location x. Similarly, $N_{iv_{z+\Delta z}}$ represents molar flow rate of component i in the gaseous phase along z direction at $z + \Delta z$. Let S_ι be the saturation of oleic phase, S_a the saturation of aqueous phase, and S_v saturation of the gaseous phase.

Now we can conserve the mass of each component (not the mass of each phase) over the elemental volume V_b and time interval of Δt such that

$$[(N_{i\iota} + N_{ia} + N_{iv})_x - (N_{i\iota} + N_{ia} + N_{iv})_{x+\Delta x}]\Delta t$$

$$+ [(N_{i\iota} + N_{ia} + N_{iv})_y - (N_{i\iota} + N_{ia} + N_{iv})_{y+\Delta y}]\Delta t$$

$$+ [(N_{i\iota} + N_{ia} + N_{iv})_z - (N_{i\iota} + N_{ia} + N_{iv})_{z+\Delta z}]\Delta t + Q_i^*\Delta t \qquad (4.57)$$

$$= [\phi\Delta x\Delta y\Delta z(S_\iota C_{i\iota} + S_a C_{ia} + S_v C_{iv})]_{t+\Delta t}$$

$$- [\phi\Delta x\Delta y\Delta z(S_\iota C_{i\iota} + S_a C_{ia} + S_v C_{iv})]_t,$$

where

Q_i^* = external injection or production of component i $\left(\dfrac{\text{moles}}{\text{day}}\right)$,

$C_{i\iota}$ = molar concentration of component i in the oleic phase $\left(\dfrac{\text{moles}}{\text{cu ft}}\right)$,

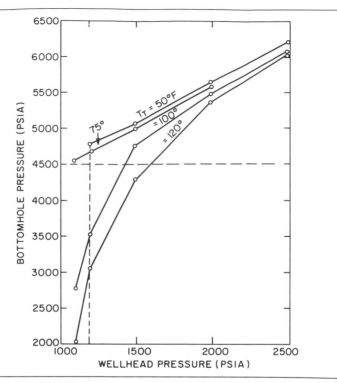

Figure 4.8 Effect of top hole injection pressure and temperature on bottom hole pressure during CO_2 injection through 2″ tubing at 9,500 feet.

C_{iv} = molar concentration of component i in the gaseous phase $\left(\dfrac{\text{moles}}{\text{cu ft}}\right)$,

C_{ia} = molar concentration of component i in the gaseous phase $\left(\dfrac{\text{moles}}{\text{cu ft}}\right)$,

$\Delta x \Delta y \Delta z$ = block dimensions in x, y, z directions (ft),

ϕ = porosity (fraction).

It is important to note that in this formulation, the term Q_i^* is to be taken as positive for injection and as negative for production.

At this point, we can introduce the molar fluxes for each of the phases (in moles per unit area per unit time):

$$
\begin{aligned}
N_{i_{v_x}} &= n_{i_{v_x}}A_x, & N_{i_{v_y}} &= n_{i_{v_y}}A_y, & N_{i_{v_z}} &= n_{i_{v_z}}A_z, \\
N_{ia_x} &= n_{ia_x}A_x, & N_{ia_y} &= n_{ia_y}A_y, & N_{ia_z} &= n_{ia_z}A_z, \\
N_{iv_x} &= n_{iv_x}A_x, & N_{iv_y} &= n_{iv_y}A_y, & N_{iv_z} &= n_{iv_z}A_z,
\end{aligned}
\tag{4.58}
$$

where $n_{i(\iota,a,v)}$ are the molar fluxes for component i in oleic, aqueous, and gaseous phases along x, y, and z directions (moles/ft²-day).

The molar flux again can be defined as

$$n_{i\iota x} = C_{i\iota}v_{\iota x}, \text{ etc.,} \tag{4.59}$$

where

$C_{i\iota}$ = molar concentration of ith component in x direction in the oleic phase (moles/ft^3),

$v_{\iota x}$ = x direction component of the velocity of oleic phase, ft/day.

It is also noted that

$$C_{i\iota} = X_{i\iota}C_{\iota}, \tag{4.60a}$$

$$C_{ia} = X_{ia}C_a, \tag{4.60b}$$

$$C_{iv} = X_{iv}C_v, \tag{4.60c}$$

where

$X_{i(\iota,a,v)}$ = mole fractions of component i in the oleic, aqueous, and gaseous phases (fraction),

$C_{(\iota,a,v)}$ = molar concentration of the oleic, aqueous, and gaseous phases, moles/ft^3, which can be obtained by dividing density of the respective phase by its molecular weight; that is,

$$C_{(\iota,a,v)} = \frac{\rho_{(\iota,a,v)}}{M_{(\iota,a,v)}}.$$

Then, one can redefine equation (4.59) as

$$n_{i\iota x} = X_{i\iota}\left(\frac{\rho_{\iota}}{N_{\iota}}\right)v_{\iota x}, \text{ etc.,} \tag{4.61}$$

and equation (4.58) can be written as

$$N_{i\iota x} = X_{ia}\left(\frac{\rho_{\iota}v_{\iota x}}{M_{\iota}}\right)A_x, \text{ etc.} \tag{4.62}$$

Substituting equation (4.62) into equation (4.57) and multiplying the first bracket by $\frac{\Delta x}{\Delta x}$, the second bracket by $\frac{\Delta y}{\Delta y}$, and the third bracket by $\frac{\Delta z}{\Delta z}$ and dividing the entire equation by Δt, one obtains

$$-\frac{\Delta x}{\Delta x}\left\{\left[\left(\frac{X_{i\iota}\rho_{\iota}v_{\iota x}}{M_{\iota}} + \frac{X_{ia}\rho_{a}v_{ax}}{M_{a}} + \frac{X_{iv}\rho_{v}v_{vx}}{M_{v}}\right)A_x\right]_{x+\Delta x}\right.$$

$$-\left[\left(\frac{X_{i\iota}\rho_\iota v_{\iota x}}{M_\iota} + \frac{X_{ia}\rho_a v_{ax}}{M_a} + \frac{X_{iv}\rho_v v_{vx}}{M_v}\right)A_x\right]_x\Biggr\}$$

$$-\frac{\Delta y}{\Delta y}\Biggl\{\left[\left(\frac{X_{i\iota}\rho_\iota v_{\iota y}}{M_\iota} + \frac{X_{ia}\rho_a v_{ay}}{M_a} + \frac{X_{iv}\rho_v v_{vy}}{M_v}\right)A_y\right]_{y+\Delta y}$$

$$-\left[\left(\frac{X_{i\iota}\rho_\iota v_{\iota y}}{M_\iota} + \frac{X_{ia}\rho_a v_{ay}}{M_a} + \frac{X_{iv}\rho_v v_{vy}}{M_v}\right)A_y\right]_y\Biggr\}$$

$$\text{(4.63)}$$

$$-\frac{\Delta z}{\Delta z}\Biggl\{\left[\left(\frac{X_{i\iota}\rho_\iota v_{\iota z}}{M_\iota} + \frac{X_{ia}\rho_a v_{az}}{M_a} + \frac{X_{iv}\rho_v v_{vz}}{M_v}\right)A_z\right]_{z+\Delta z}$$

$$-\left[\left(\frac{X_{i\iota}\rho_\iota v_{\iota z}}{M_\iota} + \frac{X_{ia}\rho_a v_{az}}{M_a} + \frac{X_{iv}\rho_v v_{vz}}{M_v}\right)A_z\right]_z\Biggr\}$$

$$+\, Q_i^* = \frac{1}{\Delta t}\Biggl\{\left[\phi V_b\left(S_\iota \frac{X_{i\iota}\rho_\iota}{M_\iota} + S_a \frac{X_{ia}\rho_a}{M_a} + S_v \frac{X_{iv}\rho_v}{M_v}\right)\right]_{t+\Delta t}$$

$$-\left[\phi V_b\left(S_\iota \frac{X_{i\iota}\rho_\iota}{M_\iota} + S_a \frac{X_{ia}\rho_a}{M_a} + S_v \frac{X_{iv}\rho_v}{M_v}\right)\right]_t\Biggr\}.$$

Taking the limits as $\Delta x \to 0$, $\Delta y \to 0$, $\Delta z \to 0$, $\Delta t \to 0$, and recalling that $\dfrac{df(x)}{dx} = f'(x) = \underset{\Delta x \to 0}{\text{Lim}}\dfrac{f(x + \Delta x) - f(x)}{\Delta x}$, one obtains

$$-\frac{\delta}{\delta x}\left(\frac{X_{i\iota}\rho_\iota v_{\iota x}}{M_\iota} + \frac{X_{ia}\rho_a v_{ax}}{M_a} + \frac{X_{iv}\rho_v v_{vx}}{M_v}\right)$$

$$-\frac{\delta}{\delta y}\left(\frac{X_{i\iota}\rho_\iota v_{\iota y}}{M_\iota} + \frac{X_{ia}\rho_a v_{ay}}{M_a} + \frac{X_{iv}\rho_v v_{vy}}{M_v}\right)$$

$$\text{(4.64)}$$

$$-\frac{\delta}{\delta z}\left(\frac{X_{i\iota}\rho_\iota v_{\iota z}}{M_\iota} + \frac{X_{ia}\rho_a v_{az}}{M_a} + \frac{X_{iv}\rho_v v_{vz}}{M_v}\right)$$

$$=\frac{\delta}{\delta t}\left[\phi V_b\left(S_\iota \frac{X_{i\iota}\rho_\iota}{M_\iota} + S_a \frac{X_{ia}\rho_a}{M_a} + S_v \frac{X_{iv}\rho_v}{M_v}\right)\right].$$

Now, the phase velocities appearing in equation (4.64) can be obtained from Darcy's equation:

$$v_{\iota x} = -(5.615)\frac{k_x k_{r\iota}}{\mu_\iota}\frac{\delta \Phi_\iota}{\delta x},$$

$$v_{ax} = -(5.615)\frac{k_x k_{ra}}{\mu_a}\frac{\delta\Phi_a}{\delta x},$$

(4.65)

$$v_{vx} = -(5.615)\frac{k_x k_{rv}}{\mu_v}\frac{\delta\Phi_v}{\delta x}.$$

Similar expressions for $v_{\iota y}$, v_{ay}, v_{vy}, $v_{\iota z}$, v_{az}, and v_{vz} can be written. In equation (4.65) the following terms are easily identified:

k_x = absolute permeability in x direction (Darcy $\cdot$ 1.127),
$k_{r(\iota,a,v)}$ = relative permeability to each phase (fraction),
$\mu_{(\iota,a,v)}$ = viscosity of each phase (cp),
$\Phi_{(\iota,a,v)}$ = potential of each phase (psia).

The potential gradient then can be defined as

$$\frac{\delta\Phi}{\delta x} = \frac{\delta p}{\delta x} - \frac{1}{144}\frac{g}{g_c}\rho\frac{\delta D}{\delta x},$$

(4.66)

where

g = local gravitational acceleration (ft/sec^2),
g_c = conversion factor lbf/lbm (32.17 ft/sec^2),
D = depth below sea level (positive downward) (ft).

Finally, substitution of Darcy's flow equation in the molar balance equation (4.63) gives

$$\frac{\delta}{\delta x}\left(\frac{X_{i\iota}\rho_\iota}{M_\iota}\frac{k_x k_{r\iota}}{\mu_\iota}A_x\frac{\delta\Phi_\iota}{\delta x} + \frac{X_{ia}\rho_a}{M_a}\frac{k_x k_{ra}}{\mu_a}A_x\frac{\delta\Phi_a}{\delta x} + \frac{X_{iv}\rho_v}{M_v}\frac{k_x k_{rv}}{\mu_v}A_x\frac{\delta\Phi_v}{\delta x}\right)\Delta x$$

$$+ \frac{\delta}{\delta y}\left(\frac{X_{i\iota}\rho_\iota}{M_\iota}\frac{k_y k_{r\iota}}{\mu_\iota}A_y\frac{\delta\Phi_\iota}{\delta y} + \frac{X_{ia}\rho_a}{M_a}\frac{k_y k_{ra}}{\mu_a}A_y\frac{\delta\Phi_a}{\delta y} + \frac{X_{iv}\rho_v}{M_v}\frac{k_y k_{rv}}{\mu_v}A_y\frac{\delta\Phi_v}{\delta y}\right)\Delta y$$

(4.67)

$$+ \frac{\delta}{\delta z}\left(\frac{X_{i\iota}\rho_\iota}{M_\iota}\frac{k_z k_{r\iota}}{\mu_\iota}A_z\frac{\delta\Phi_\iota}{\delta z} + \frac{X_{ia}\rho_a}{M_a}\frac{k_z k_{ra}}{\mu_a}A_z\frac{\delta\Phi_a}{\delta z} + \frac{X_{iv}\rho_v}{M_v}\frac{k_z k_{rv}}{\mu_v}A_z\frac{\delta\Phi_v}{\delta_z}\right)\Delta z + \frac{Q_i^*}{5.615}$$

$$= \frac{V_b}{5.615}\frac{\delta}{\delta t}\left(\frac{X_{i\iota}S_\iota\rho_\iota\phi}{M_\iota} + \frac{X_{ia}S_a\rho_a\phi}{M_a} + \frac{X_{iv}S_v\rho_v\phi}{M_v}\right).$$

At this point, we note that we can no longer say that the mass of each phase is conserved because of the transfer of various components between the phases. This development allows not only the hydrocarbon component transfer between the oil and gas phases but also component transfer between the gas phase and water phase (such as CO_2). Finally, this development will allow the vaporization of water into the gas phase (as happens in steam flooding).

It is obvious that for an N component system there will be N differential equations [eq. (4.67)]. But a close look at equation (4.62) reveals the fact that there are many more dependent variables of the problem. The dependent variables (unknowns) of the system are summarized in table 4.3. As we can see from the table, there are $3N + 6$ unknowns. It should also be remembered that some other properties such as densities $(\rho_\iota, \rho_a, \rho_v)$, viscosities $(\mu_\iota, \mu_a, \mu_v)$, and relative permeabilities $(k_{r\iota}, k_{ra}, k_{rv})$ can be treated as unknowns since they are functions of the phase pressures, phase compositions, and saturations. The functional relationships that define the densities and viscosities as functions of compositions and pressures and relative permeabilities as functions of saturations are readily available from fluid and rock data. Therefore, we focus our attention to $3N + 6$ unknowns. In order to solve this system uniquely, we must have $3N + 6$ independent relationships. The listing of the relationships is given in table 4.4. The algebraic equations that appear in table 4.4 are described explicitly as follows.

Phase equilibriums: For each pair of phases there is an equilibrium coefficient for each component, such as

$$\frac{X_{iv}}{X_{i\iota}} = K_{i\iota v} \qquad i = 1,2, \ldots ,N;$$

$$\frac{X_{iv}}{X_{ia}} = K_{iva} \qquad i = 1,2, \ldots ,N; \qquad (4.68)$$

$$\frac{X_{ia}}{X_{i\iota}} = K_{ia\iota} \qquad i = 1,2, \ldots ,N.$$

The equilibrium coefficients may be known functions of pressure and composition, or they can be the solution of a nonlinear set of equations derived from thermodynamic principles. Equation (4.68) provides $3N$ equations when written for each component in the system. However, the third equation is not independent of the first two since it is derived from the others.

Capillary pressure: There are two independent capillary pressure relationships:

$$P_{cv\iota}(S_\iota) = P_v - P_\iota$$

Table 4.3 Unknowns of the N Component System

Variables	Number of Unknowns
$X_{i\iota}$	N
X_{ia}	N
X_{iv}	N
P_ι, P_a, P_v	3
S_ι, S_a, S_v	3
Total	$3N + 6$

Table 4.4 Listing of Relationships

Relationship	Type of Equation	Number of Equations
Component balances	Differential	N
Phase equilibriums	Algebraic	$2N$
Capillary pressure	Algebraic	2
Saturation	Algebraic	1
Mass fractions	Algebraic	3
Total		$3N + 6$

and

$$P_{cua}(S_a) = P_\iota - P_a. \tag{4.69}$$

These capillary pressure relationships assume that gas-phase pressure is greater than oil-phase pressure, which is in turn greater than water-phase pressure (water-wet system).

Saturation: The phase saturations must always sum to unity since the pores are always 100% fluid filled:

$$S_\iota + S_a + S_v = 1. \tag{4.70}$$

Mass fraction: In each phase, mass fractions of each component must add up to 1 since mass conservation of each component is required:

$$\sum_{i=1}^{N} X_{i\iota} = 1,$$

$$\sum_{i=1}^{N} X_{ia} = 1, \tag{4.71}$$

$$\sum_{i=1}^{N} X_{iv} = 1.$$

4.8 **APPENDIX: DERIVATION OF CONTINUITY EQUATION FOR BLACK OIL MODELING.** The compositional model developed in the previous section is extremely complex to set up and solve. A black oil model is a simplified version of a compositional model. In fact, black oil modeling is a subset of compositional modeling. A black oil model assumes that three components exist in the system: oil, gas, and water. Gas may dissolve in oil and to a certain extent in water, but neither oil nor water vaporizes into the gas phase. The description of this phase behavior is input directly to the model as adjusted differential liberation data. The basic assumption of the black oil model is that

Darcy's Law (as modified by relative permeability to account for multiphase flow) adequately describes the mass-transport phenomenon. Dispersion is not included in the model formulation.

In the following section we develop the black oil model equations, starting with the general equation of the compositional model, equation (4.67). We start with three components and three phases:

$i = 1$ oil component, ι = oleic phase,
$i = 2$ water component, a = aqueous phase,
$i = 3$ gas component, v = gaseous phase.

Furthermore, it is usual to assume that the oil component can exist only in the oil phase, that the water component can exist only in the aqueous phase, and that the gas component can exist in all three phases. Thus,

$$X_{1\iota} + X_{3\iota} = 1, \qquad (X_{2\iota} = 0, X_{1\iota} \neq 0),$$

$$X_{2a} + X_{3a} = 1, \qquad (X_{1a} = 0, X_{2a} \neq 0), \qquad\qquad (4.72)$$

$$X_{3v} = 1, \qquad (X_{1v} = 0, X_{2v} = 0).$$

Gas solubility (also called solution gas-oil ratio, R_{so}, or solution gas-water ratio, R_{sw}) is defined as the volume of gas (SCF measured at standard conditions) dissolved at a given temperature and pressure in one STB of oil or water. Then,

$$X_{3\iota} = \frac{\text{moles of gas}}{\text{total moles of oleic phase}} = \frac{\dfrac{R_{so}}{379.1}}{\dfrac{(5.615)(\rho_\iota)(B_\iota)}{M_\iota}} = \frac{R_{so}M_\iota}{(5.615)(\rho_\iota)(B_\iota)(379.1)},$$

$$X_{1\iota} = \frac{\text{moles of oil}}{\text{total moles of oleic phase}} = \frac{\dfrac{(1\ \text{STB})(5.615)(\rho_{osc})}{M_o}}{\dfrac{(5.615)(\rho_\iota)(B_\iota)}{M_\iota}} = \frac{\rho_{osc}M_\iota}{M_o\rho_\iota B_\iota},$$

$$X_{3a} = \frac{\text{moles of gas}}{\text{total moles of aqueous phase}} = \frac{\dfrac{R_{sw}}{379.1}}{\dfrac{(5.615)(\rho_a)(B_a)}{M_a}} = \frac{R_{sw}M_a}{(5.615)(\rho_a)(B_a)(379.1)},$$

$$X_{2a} = \frac{\text{moles of water}}{\text{total moles of aqueous phase}} = \frac{\dfrac{(1\ \text{STB})(5.615)(\rho_{wsc})}{M_w}}{\dfrac{(5.615)(\rho_a)(B_a)}{M_a}} = \frac{\rho_{wsc}M_a}{M_w\rho_a B_a},$$

where M_ι and M_a are the molecular weights of the oleic and aqueous phases respectively. At this stage we can rewrite equation (4.67) for components $i=1$ (oil), $i=2$ (water), and $i=3$ (gas) by plugging the definitions of these mole fractions.

$i=1$ **Oil Equation.** Substituting for $X_{1\iota}$ and observing that $X_{1a}=0$, $X_{1v}=0$, and setting $\iota \equiv o$, one obtains the oil equation as follows:

$$\frac{\delta}{\delta x}\left(\frac{A_x k_x k_{ro}}{\mu_o B_o}\frac{\delta \Phi_o}{\delta x}\right)\Delta x + \frac{\delta}{\delta y}\left(\frac{A_y k_y k_{ro}}{\mu_o B_o}\frac{\delta \Phi_o}{\delta y}\right)\Delta y$$

$$+ \frac{\delta}{\delta z}\left(\frac{A_z k_z k_{ro}}{\mu_o B_o}\frac{\delta \Phi_o}{\delta z}\right)\Delta z + q_o = \frac{V_b}{5.615}\frac{\delta}{\delta t}\left(\frac{\phi S_o}{B_o}\right). \tag{4.73}$$

It should be noted that Q_1^* of equation (4.67) is also replaced by

$$Q_1^* = \frac{(5.615)\left(q_o\dfrac{STB}{D}\right)(\rho_{osc})}{M_o}.$$

$i=2$ **Water Equation.** Substituting for X_{2a} and observing that $X_{2\iota}=0$, $X_{2v}=0$, and setting $a \equiv w$, one can obtain the water equation in a similar manner:

$$\frac{\delta}{\delta x}\left(\frac{A_x k_x k_{rw}}{\mu_w B_w}\frac{\delta \Phi_w}{\delta_x}\right)\Delta x + \frac{\delta}{\delta y}\left(\frac{A_y k_y k_{rw}}{\mu_w B_w}\frac{\delta \Phi_w}{\delta_y}\right)\Delta y$$

$$+ \frac{\delta}{\delta z}\left(\frac{A_z k_z k_{rw}}{\mu_w B_w}\frac{\delta \Phi_w}{\delta z}\right)\Delta z + q_w = \frac{V_b}{5.615}\frac{\delta}{\delta t}\left(\frac{\phi S_w}{B_w}\right). \tag{4.74}$$

Again, note that Q_2^* of equation (4.67) is replaced by

$$Q_2^* = \frac{(5.615)\left(q_w\dfrac{STB}{D}\right)(\rho_{wsc})}{M_w}.$$

$i=3$ **Gas Equation.** This time we substitute for $X_{3\iota}$ and X_{3a} and observe that $X_{1v}=0$, $X_{2v}=0$, and $X_{3v}=1$. Then, the gas equation can be obtained as follows ($v \equiv g$):

$$\frac{\delta}{\delta x}\left(R_{so}\frac{A_x k_x k_{ro}}{B_o \mu_o}\frac{\delta \Phi_o}{\delta x} + R_{sw}\frac{A_x k_x k_{rw}}{B_w \mu_w}\frac{\delta \Phi_w}{\delta x} + \frac{A_x k_x k_{rg}}{B_g \mu_g}\frac{\delta \Phi_g}{\delta x}\right)\Delta x$$

$$+ \frac{\delta}{\delta y}\left(R_{so}\frac{A_y k_y k_{ro}}{B_o \mu_o}\frac{\delta \Phi_o}{\delta y} + R_{sw}\frac{A_y k_y k_{rw}}{B_w \mu_w}\frac{\delta \Phi_w}{\delta y} + \frac{A_y k_y k_{rg}}{B_g \mu_g}\frac{\delta \Phi_g}{\delta y}\right)\Delta y$$

(4.75)

$$+ \frac{\delta}{\delta z}\left(R_{so}\frac{A_z k_z k_{ro}}{B_o \mu_o}\frac{\delta \Phi_o}{\delta z} + R_{sw}\frac{A_z k_z k_{rw}}{B_w \mu_w}\frac{\delta \Phi_w}{\delta z} + \frac{A_z k_z k_{rg}}{B_g \mu_g}\frac{\delta \Phi_g}{\delta z}\right)\Delta z$$

$$+ q_g = \frac{V_b}{5.615}\frac{\delta}{\delta t}\left(R_{so}\frac{\phi S_o}{B_o} + R_{sw}\frac{\phi S_w}{B_w} + \frac{\phi S_g}{B_g}\right).$$

In the development of equation (4.75), Q_3^* of equation (4.67) is replaced by

$$Q_3^* = q_g\left(\frac{\text{SCF}}{D}\right)\left(\frac{1}{379.1}\right),$$

and B_g is defined by

$$B_g = \frac{(P_{sc})(z)(T)}{(5.615)(T_{sc})(P)}\left(\frac{\text{bbl}}{\text{SCF}}\right).$$

Examination of equation (4.75) clearly indicates that on the left-hand side along each flow direction, transportation of gas (component 3) in each phase (oleic, aqueous, and gaseous) is accommodated. Similarly, on the right-hand side of the equation (accumulation terms), accumulation of gas in all three phases is allowed.

As can be seen from equations (4.73), (4.74), and (4.75) in the black oil model, we have a total of only six unknowns—namely, phase saturations (S_o, S_w, and S_g) and phase pressures. Therefore, we need three more equations to solve the system uniquely. These equations are again readily available from capillary pressure relationships; that is,

$$P_{cow}(S_w) = P_o - P_w$$

and

$$P_{cgo}(S_g) = P_g - P_o,$$

(4.76)

and the saturation relationship

$$S_o + S_w + S_g = 1.$$

(4.77)

4.9   APPENDIX: FINITE DIFFERENCE APPROXIMATIONS OF FLOW EQUATIONS AND THEIR SOLUTION. The purpose of finite difference approximations is to express a differential equation as an algebraic equation approximating the original equation at a

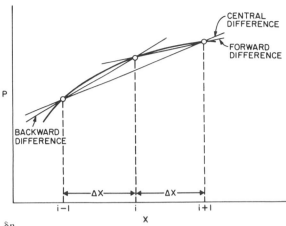

First Derivative: $\dfrac{\delta p}{\delta x}$

$$\frac{\delta p}{\delta x} = \frac{p_{i+1} - p_i}{\Delta x} \qquad \text{Forward difference,}$$

$$\frac{\delta p}{\delta x} = \frac{p_i - p_{i-1}}{\Delta x} \qquad \text{Backward difference,}$$

$$\frac{\delta p}{\delta x} = \frac{p_{i+1} - p_{i-1}}{2\Delta x} \qquad \text{Central difference.}$$

Second Derivative: $\dfrac{\delta^2 p}{\delta x^2}$

$$\frac{\delta^2 p}{\delta x^2} = \frac{\left(\dfrac{\delta p}{\delta x}\right)_{i+1/2} - \left(\dfrac{\delta p}{\delta x}\right)_{i-1/2}}{\Delta x} = \frac{\dfrac{p_{i+1} - p_i}{\Delta x} - \dfrac{p_i - p_{i-1}}{\Delta x}}{\Delta x}$$

$$= \frac{p_{i+1} - 2p_i + p_{i-1}}{(\Delta x)^2}.$$

Figure 4.9 Finite difference approximations.

specific point. For instance, if the value of pressure, p, is represented graphically as a function of distance, x, as shown on figure 4.9, then the derivative of p with respect to x at x_i may be approximated in several ways using the values of the pressure p at known points in the immediate vicinity of i. Each of the three approximations to the first derivative shown on figure 4.9 is equally valid.

The second derivative of pressure, with respect to distance, can be determined in a similar manner, but in this instance, adjacent values of the first derivative are used as a function of distance. If the approximations to the first derivatives are used rather than the exact values of these derivatives, we obtain an approximation to the second derivative of pressure with respect to distance in terms of three pressure values as shown on figure 4.9.

Taylor's Series Approximation of p_{i+1}:

$$p_{i+1} = p_i + \Delta x \left(\frac{\delta p}{\delta x}\right)_i + \frac{(\Delta x)^2}{2}\left(\frac{\delta^2 p}{\delta x^2}\right)_i + \frac{(\Delta x)^3}{3!}\left(\frac{\delta^3 p}{\delta x^3}\right)_i + \cdots \tag{1}$$

Difference Approximation to First Derivative:

$$\left(\frac{\delta p}{\delta x}\right)_i = \frac{p_{i+1} - p_i}{\Delta x} - \frac{(\Delta x)}{2}\left(\frac{\delta^2 p}{\delta x^2}\right)_i - \frac{(\Delta x)^2}{3!}\left(\frac{\delta^3 p}{\delta x^3}\right) - \cdots \tag{2}$$

Second Derivative Approximation:

Taylor's Series Approximation of p_{i-1}:

$$p_{i-1} = p_i - \Delta x \left(\frac{\delta p}{\delta x}\right)_i + \frac{(\Delta x)^2}{2}\left(\frac{\delta^2 p}{\delta x^2}\right)_i - \frac{(\Delta x)^3}{3!}\left(\frac{\delta^3 p}{\delta x^3}\right)_i + \cdots \tag{3}$$

Combining equations (1) and (3):

$$\left(\frac{\delta^2 p}{\delta x^2}\right)_i = \frac{p_{i+1} - 2p_i + p_{i-1}}{(\Delta x)^2} - \frac{(\Delta x)^2}{12}\left(\frac{\delta^4 p}{\delta x^4}\right)_i - \cdots \tag{4}$$

Higher Order Approximation:

$$\left(\frac{\delta^2 p}{\delta x^2}\right)_i = \frac{-p_{i+2} + 16p_{i+1} - 30p_i + 16p_{i-1} - p_{i-2}}{12(\Delta x)^2} + O(\Delta x)^4 \tag{5}$$

Figure 4.10 Taylor series and truncation error.

The errors associated with these derivative approximations are referred to as *truncation* errors and can be quantified through the use of Taylor's series. As shown on figure 4.10, the expansion of a Taylor's series about a point i allows us to estimate as close as we like the function value of an adjacent point $i + 1$, provided we know the values of the function derivatives at the point i.

The terms containing derivatives that remain in the series are referred to as the truncation error because they are dropped, or truncated, from the approximation. The truncation error is referred to as of *order* Δx or *order* $(\Delta x)^2$, depending on the form of the first term in the truncated portion of the series. By combining the Taylor's series approximations to the $i + 1$ and $i - 1$ points, we obtain an approximation to the second derivative and determine that the truncation error is of order $(\Delta x)^2$ as shown on figure 4.10.

By successive use of Taylor's series at points more distant than $i + 1$ and $i - 1$ (say, $i + 2$ and $i - 2$), we are able to develop higher order approximations to the derivatives at point i as illustrated at the bottom of figure 4.10. This approximation was obtained by algebraically manipulating the Taylor's series expansion to points $i + 2$, $i + i$, $i - 1$, and $i - 2$ to eliminate the lower order terms and arrive at the approximate expression to the second derivative.

Using the finite difference approximations just discussed, one may, for example, rewrite the single-phase flow in porous media (diffusivity) equation (fig. 4.11) in finite

General Form:

$$\frac{\delta}{\delta x}\left(\frac{k_x}{\mu}\frac{\delta p}{\delta x}\right) + \frac{\delta}{\delta y}\left(\frac{k_y}{\mu}\frac{\delta p}{\delta y}\right) - \frac{q_v}{\rho} = (c_r + c_f)\phi\frac{\delta p}{\delta t}.$$

If $k_x = k_y$ and is constant together with μ across the reservoir, then we obtain the standard form:

$$\frac{\delta^2 p}{\delta x^2} + \frac{\delta^2 p}{\delta y^2} - \frac{\mu}{k\rho}q_v = \frac{c_T\phi\mu}{k}\frac{\delta p}{\delta t}$$

Figure 4.11 Diffusivity equation.

$$\left[\frac{p_{i+1,j} - 2p_{i,j} + p_{i-1,j}}{(\Delta x)^2} + \frac{P_{i,j+1} - 2p_{i,j} + p_{i,j-1}}{(\Delta y)^2} - \left(\frac{\mu}{kh}q_v\right)_{ij}\right]_{n\ or\ n+1}$$

$$= \frac{c_T\phi\mu}{kh}\left(\frac{p_{n+1} - p_n}{\Delta t}\right)_{i,j}.$$

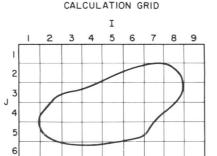

CALCULATION GRID

Figure 4.12 Two-dimensional diffusivity equation in finite difference form.

difference form as shown in figure 4.12. Also shown in figure 4.12 is an example of a calculation mesh or grid that is used to locate those discrete points in the reservoir for which the difference equation can be written.

Note that we have an option when writing the spatial derivatives—namely, $\frac{\delta^2 P}{\delta x^2}$ and $\frac{\delta^2 P}{\delta y^2}$—of expressing the time level at the current time level, n, or at the next time level, $n + 1$. If we choose the n time level, we have an *explicit* expression, while a choice of time level $n + 1$ is called *implicit*. For this equation, time level $n + 1$ is virtually always chosen to assure that the resulting computations are stable.

Writing the algebraic difference equations for each grid block results in an interdependent set of equations that can be expressed in the form of a matrix as shown in figure 4.13. These equations may be solved by any of a variety of techniques including PSOR, LSOR, SIP, ADIP, and direct solution. (Peaceman and Rachford 1955; Stone

Calculation Grid:
$$I =$$

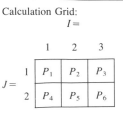

Linear Systems of Equations:

$$c_1 p_1 + d_1 p_2 \qquad + e_1 p_4 \qquad\qquad\qquad = f_1$$

$$b_2 p_1 + c_2 p_2 + d_2 p_3 \qquad + e_2 p_5 \qquad\qquad = f_2$$

$$\qquad + b_3 p_2 + c_3 p_3 + d_3 p_4 \qquad + e_3 p_6 = f_3$$

$$a_4 p_1 \qquad\qquad + b_4 p_3 + c_4 p_4 + d_4 p_5 \qquad = f_4$$

$$\qquad a_5 p_2 \qquad\qquad + b_5 p_4 + c_5 p_5 + d_5 p_6 = f_5$$

$$\qquad\qquad a_6 p_3 \qquad\qquad + b_6 p_5 + c_6 p_6 = f_6$$

Matrix Equation:

$$
\begin{bmatrix}
c_1 & d_1 & 0 & e_1 & 0 & 0 \\
b_2 & c_2 & d_2 & 0 & e_2 & 0 \\
0 & b_3 & c_3 & d_3 & 0 & e_3 \\
a_4 & 0 & b_4 & c_4 & d_4 & 0 \\
0 & a_5 & 0 & b_5 & c_5 & d_5 \\
0 & 0 & a_6 & 0 & b_6 & c_6
\end{bmatrix}
\begin{bmatrix}
P_1 \\ P_2 \\ P_3 \\ P_4 \\ P_5 \\ P_6
\end{bmatrix}
=
\begin{bmatrix}
f_1 \\ f_2 \\ f_3 \\ f_4 \\ f_5 \\ f_6
\end{bmatrix}
$$

Figure 4.13 System of simulation equations.

1968; Varga 1962). Many of these are given as options in most reservoir simulation codes.

Usually, the more typical form of the diffusivity equation would be used in a simulation model as shown in figure 4.12. Using the difference representation at discrete points across the reservoir allows us to vary the permeability and porosity across the grid. This flexibility is one of the main advantages to using numerical techniques in the analysis of reservoir engineering problems.

Figure 4.14 also presents definitions of *transmissibility* and *pore volume*, which are common terms in the reservoir simulation vernacular. Pore volume is self-explanatory. The production term is a volumetric flow rate. Transmissibility by definition is nothing more than the Darcy's Law coefficient for flow between two adjacent grid blocks. In order to calculate the transmissibility, however, it is necessary to define the interblock value of both permeability and viscosity. If we carry the Darcy's Law basis for the definition of the transmissibility one step further, we realize that the proper average for the permeability/viscosity ratio is a harmonic average of the properties in each block since the flow is in series.

Differential Equation:

$$\frac{\delta}{\delta x}\left(\frac{k_x h}{\mu}\frac{\delta p}{\delta x}\right) + \frac{\delta}{\delta y}\left(\frac{k_y h}{\mu}\frac{\delta p}{\delta y}\right) - \frac{q_v h}{\rho} = (c_r + c_f)\,\phi\,h\,\frac{\delta p}{\delta t}.$$

Difference Form:

$$\frac{1}{(\Delta x)^2}\left[\left(\frac{k_x h}{\mu}\right)_{i+1/2,j}(p_{i+1,j} - p_{i,j}) - \left(\frac{k_x h}{\mu}\right)_{i-1/2,j}(p_{i,j} - p_{i-1,j})\right]^{n+1}$$

$$+ \frac{1}{(\Delta y)^2}\left[\left(\frac{k_y h}{\mu}\right)_{i,j+1/2}(p_{i,j+1} - p_{i,j}) - \left(\frac{k_y h}{\mu}\right)_{i,j-1/2}(p_{i,j} - p_{i,j-1})\right]^{n+1} - \left(\frac{q_v h}{\rho}\right)_{i,j}$$

$$= (c_r + c_f)\,\phi\,h\left(\frac{p_{n+1} - p_n}{\Delta t}\right)_{i,j}$$

or

$$\left[\left(\frac{k_x h(\Delta y)}{\mu(\Delta x)}\right)_{i+1/2,j}(p_{i+1,j} - p_{i,j}) - \left(\frac{k_x h(\Delta y)}{\mu(\Delta x)}\right)_{i-1/2,j}(p_{i,j} - p_{i-1,j})\right]^{n+1}$$

$$+ \left[\left(\frac{k_y h(\Delta x)}{\mu(\Delta y)}\right)_{i,j+1/2}(p_{i,j+1} - p_{i,j}) - \left(\frac{k_y h(\Delta x)}{\mu(\Delta y)}\right)_{i,j-1/2}(p_{i,j} - p_{i,j-1})\right]^{n+1} - \left(\frac{q_v \Delta x \Delta y h}{\rho}\right)_{i,j}$$

$$= (c_r + c_f)\,\phi\Delta x\,\Delta y h\left(\frac{p_{n+1} - p_n}{\Delta t}\right)_{i,j}.$$

Define:

Transmissibility $\quad T_{x_{i+1/2,j}} = \left[\dfrac{k_x h(\Delta y)}{\mu(\Delta x)}\right]_{i+1/2,j},$

Pore volume $\quad V_P = (\phi\Delta x\,\Delta y h)_{i,j},$

Production $\quad Q = \left(\dfrac{q_v \Delta x \Delta y h}{\rho}\right)_{i,j}.$

The diffusivity equation is

$$[T_{x_{i+1/2,j}}(p_{i+1,j} - p_{i,j}) - T_{x_{i-1/2,j}}(p_{i,j} - p_{i-1,j})]^{n+1}$$

$$+ [T_{y_{i,j+1/2}}(p_{i,j+1} - p_{i,j}) - T_{y_{i,j-1/2}}(p_{i,j} - p_{i,j-1})]^{n+1} - Q$$

$$= (c_r + c_f)\,V_P\left(\frac{p_{n+1} - p_n}{\Delta t}\right)_{i,j}$$

Figure 4.14 Two-dimensional single-phase diffusivity equation.

REFERENCES

Aziz, K., Govier, G. W., and Forgarasi, M.: "Pressure Drop in Wells Producing Oil and Gas," *J. Can. Pet. Tech.* (July–Sept., 1972) 38–48.

Beggs, H. D., and Brill, J. P.: "A Study of Two-Phase Flow in Inclined Pipes," *J. Pet. Tech.* (May 1973) 608–617.

Bilhartz, H. L., Charlson, G. S., Stalkup, F. I., and Miller, C. C.: "A Method for Projecting Full-Scale Performance of CO_2 Flooding in the Willard Unit," paper SPE 7051 presented at the 5th SPE Symposium on Enhanced Oil Recovery, Tulsa, April 16–19, 1978.

Blackwell, R. J., Rayne, J. R., and Terry, W. M.: "Factors Influencing the Efficiency of Miscible Displacement," *Trans., AIME* (1959) **216,** 1–8.

Chase, C. A., Jr., and Todd, M. R.: "Numerical Simulation of CO_2 Flood Performance," paper SPE 10514 presented at the 6th SPE Symposium on Reservoir Simulation, New Orleans, January 31–February 3, 1982.

Chaudhari, N. M.: "An Improved Numerical Technique for Solving Multidimensional Miscible Displacement Equations," *Soc. Pet. Eng. J.* (Sept. 1971) 277–284.

Chierici, G. L., Schlocchi, G., and Terzi, L.: "Pressure Temperature Profiles Are Calculated for Gas Flow," *Oil and Gas J.* (Jan. 7, 1980) 65–72.

Coats, K. H., "An Equation of State Compositional Model," *Soc. Pet. Eng. J.* (Oct. 1980) 363–376.

Colebrook, C. F.: "Turbulent Flow in Pipes with Particular References to the Transition Region Between the Smooth and Rough Pipe Laws," *Ind. and Eng. Chem.* (Nov. 1939) 133.

Compositional Reservoir Simulation, Intercomp, Houston (1981) (short course notes).

Cronshaw, M. B., and Bolling, J. D.: "A Numerical Model of the Non-Isothermal Flow of Carbon Dioxide in Wellbores," paper SPE 10735 presented at the 1982 California Regional Meeting of SPE, San Francisco, March 24–26, 1982.

Desch, J. B., Larsen, W. K., Lindsay, R. F., and Nettle, R. L.: "Enhanced Oil Recovery by CO_2 Miscible Displacement in the Little Knife Field, Billings County, North Dakota," paper SPE/DOE 10696 presented at the 3rd SPE/DOE Symposium on Enhanced Oil Recovery, Tulsa, April 4–7, 1982.

Douglas, J., Jr., Peaceman, D. W., and Rachford, H. H., Jr.: "A Method for Calculating Multidimensional Immiscible Displacement," *Trans., AIME* (1959) **216,** 297–306.

Duns, H., and Ros, N. C.: "Vertical Flow of Gas and Liquid Mixtures in Wells," *Proc.,* Sixth World Pet. Cong., Frankfurt (1963) **694,** 10.

Fayers, F. J., Hawes, R. I., and Mathews, J. D.: "Some Aspects of the Potential Application of Surfactants or CO_2 as EOR Processes in North Sea Reservoirs," *J. Pet. Tech.* (Sept. 1981) 1617–1627.

Fussell, L. T., and Fussell, D. D.: "An Iterative Technique for Compositional Reservoir Models," *Soc. Pet. Eng. J.* (Aug. 1979) 211–220.

Gardner, A. O., Peaceman, D. W., and Pozzi, A. L.: "Numerical Calculation of Multidimensional Miscible Displacement by Method of Characteristics," *Soc. Pet. Eng. J.* (March 1964) 26–36.

Gardner, J. W., and Ypma, J. G. J.: "An Investigation of Phase Behavior–Macroscopic Bypassing Interaction in CO_2 Flooding," paper SPE/DOE 10686 presented at the SPE/DOE 3rd Joint Symposium on Enhanced Oil Recovery, Tulsa, April 4–7, 1982.

Gould, T. L.: "Vertical Two-Phase Steam-Water Flow in Geothermal Wells," *J. Pet. Tech.* (Aug. 1974) 833–842.

Graue, D. J., and Zana, E. T.: "Study of a Possible CO_2 Flood in Rangely Field," *J. Pet. Tech.* (July 1981) 1312–1318.

Griffith, P., and Wallis, G. B.: "Two-Phase Slug Flow," *J. Heat Trans.* (Aug. 1961) 307–320.

Hagedorn, A. R., and Brown, K. E.: "Experimental Study of Pressure Gradients Occurring During Continuous Two-Phase Flow in Small Diameter Vertical Conduits," *J. Pet. Tech.* (April 1965) 694.

Kazemi, H., Vestal, C. R., and Shank, G. D.: "An Efficient Multicomponent Numerical Simulator," *Soc. Pet. Eng. J.* (Oct. 1978) 355–368.

Klins, M. A., and Farouq Ali, S. M.: "Heavy Oil Production by Carbon Dioxide Injection," *J. Can. Pet. Tech.* (Sept.–Oct. 1982) 64–72.

Ko, S. C. M.: "A Preliminary Study of CO_2 Flooding in a Watered-out Reservoir in Alberta, Canada," paper CIM 82–33–76 presented at the 33rd Annual Technical Meeting of the CIM, Calgary, June 6–9, 1982.

Koval, E. J.: "A Method for Predicting the Performance of Unstable Miscible Displacement in Heterogeneous Media," *Soc. Pet. Eng. J.* (June 1963) 145–154.

Lacey, J. W., Faris, J. E., and Brinkman, F. H.: "Effect of Bank Size on Oil Recovery in the High-Pressure Gas-Driven LPG-Bank Process," *J. Pet. Tech.* (Aug. 1961) 806–816.

Lantz, R. B.: "Rigorous Calculation of Miscible Displacement Using Immiscible Reservoir Simulators," *Soc. Pet. Eng. J.* (June 1970) 192–202.

Larson, R. G.: "Controlling Numerical Dispersion by Timed Flux Updating in One Dimension," *Soc. Pet. Eng. J.* (June 1982) 399–408.

Muskat, M.: *The Flow of Homogeneous Fluids Through Porous Media,* McGraw-Hill, New York (1937).

Nghiem, L. X., Fong, D. K., and Aziz, K.: "Compositional Modeling with an Equation of State," *Soc. Pet. Eng. J.* (Dec. 1981) 687–698.

Nolen, J. S.: "Numerical Simulation of Compositional Phenomena in Petroleum Reservoirs," paper SPE 4274 presented at the 3rd SPE Symposium on Numerical Simulation of Reservoir Performance, Houston, January 11–12, 1973.

Orkiszewski, J.: "Predicting Two-Phase Pressure Drops in Vertical Pipe," *J. Pet. Tech.* (June 1967) 829–838.

Patton, J. T., Coats, K. H., and Spence, K.: "Carbon Dioxide Well Stimulation: Part 1—A Parametric Study," *J. Pet. Tech.* (Aug. 1982a) 1798–1804.

Patton, J. T., Sigmund, P., Evans, B., Ghose, S., and Weinbrandt, D.: "Carbon Dioxide Well Stimulation: Part 2—Design of Aminoil's North Bolsa Strip Project," *J. Pet. Tech.* (Aug. 1982b) 1805–1810.

Paul, G. W., Ramesh, B., and Gould, T. L.: *Advanced Reservoir Engineering,* Intercomp, Houston (1980).

Peaceman, D. W., and Rachford, H. H.: "The Numerical Solution of Parabolic and Elliptic Differential Equations," *J. Soc. Indust. Appl. Math* (1955) **3**, 28–41.

Peaceman, D. W., and Rachford, H. H., Jr.: "Numerical Calculation of Multidimensional Miscible Displacement," *Soc. Pet. Eng. J.* (Dec. 1962) 327–339.

Perrine, R. L., and Gay, G. M.: "Unstable Miscible Flow in Heterogeneous Systems," *Soc. Pet. Eng. J.* (Sept. 1966) 228–238.

Pontious, S. B., and Tham, M. J.: "North Cross (Devonian Unit CO_2 Flood)," *J. Pet. Tech.* (Dec. 1978) 1706–1714.

Price, H. S., and Donahue, D. A. T.: "Isothermal Displacement Processes with Interphase Mass Transfer," *Soc. Pet. Eng. J.* (June 1967) 205–220.

Reid, T. B., and Robinson, H. J.: "Lick Creek Meakin Sand Unit Immiscible CO_2—Waterflood Project," *J. Pet. Tech.* (Sept. 1981) 1723–1729.

Spivak, A., Perryman, T. L., and Norris, R. A.: "A Compositional Simulation Study of the SACROC Unit CO_2 Project," paper presented at the 9th World Pet. Cong., Tokyo, May 11–16, 1975.

Stone, H. L.: "Iterative Solutions of Implicit Approximations of Multidimensional Partial Differential Equations," *SIAM J. on Num. Anal.* (1968) **5**, 530–558.

Stright, D. H., Jr., Aziz, K., Settari, A., and Starratt, F. E.: "Carbon Dioxide Injection into Bottom-Water, Undersaturated Viscous Oil Reservoirs," *J. Pet. Tech.* (Oct. 1977) 1248–1258.

Todd, M. R., Cobb, W. M., and McCarter, E. D.: "CO_2 Flood Performance Evaluation for the Cornell Unit, San Andres Field," *J. Pet. Tech.* (Oct. 1982) 2271–2282.

Todd, M. R., and Longstaff, W. J.: "The Development, Testing and Application of a Numerical Simulator for Predicting Miscible Flood Performance," *J. Pet. Tech.* (July 1972) 874.

Varga, R. S.: *Matrix Iterative Analysis,* Prentice-Hall, Englewood Cliffs, N.J. (1962).

Warner, H. R., Jr.: "An Evaluation of Miscible CO_2 Flooding in Waterflooded Sandstone Reservoirs," *J. Pet. Tech.* (Oct. 1977) 1339–1347.

West, W. J., Garvin, W. W., and Sheldon, J. W.: "Solution of the Equations of Unsteady State Two-Phase Flow in Oil Reservoirs," *Trans., AIME* (1954) **201,** 217–229.

Willhite, G. P.: "Over-all Heat Transfer Coefficients in Steam and Hot Water Injection Wells," *J. Pet. Tech.* (May 1967) 607–615.

Youngren, G. K., and Charlson, G. S.: "History Match Analysis of the Little Creek CO_2 Pilot Test," *J. Pet. Tech.* (Nov. 1980) 2042–2052.

5 FIELD SCREENING FOR APPLICATIONS

Earlier chapters have systematically developed a discussion of how oil is displaced in the reservoir, how CO_2 injection augments those displacement mechanisms, and how one adequately predicts a reservoir's response through detailed numerical simulation. Chapter 5 is intended to bridge the gap between the technical and economic aspects of CO_2 flooding through a discussion of field screening for applications.

Essentially, the design of an EOR project is a multiple-hurdle decision-making task. Five phases are usually involved, as shown on the decision tree of figure 5.1. Phase 1 encompasses the data-gathering process, including lab and field measurements, production data, and geologic information. It also includes the identification of critical reservoir parameters and preliminary screening of the field for potential EOR applications. Phase 2 then involves reservoir response predictions and economic evaluation. Phase 3 incorporates pilot test design and implementation and re-evaluation of the proposed process in a prospective reservoir as a result of the pilot test. Phase 4 represents the final design and development of the project, and Phase 5 involves the project's later commercial operation.

This chapter focuses on the first phase, involving economic screening of a field or reservoir for application of CO_2 injection for EOR. There are fundamentally two approaches to screening: (1) binary evaluation criteria and (2) project economic criteria.

The first method, binary evaluation, requires a review of key reservoir parameters with respect to observed characteristics of technically and economically successful CO_2 floods. Due to the limited number of projects and amount of published data, this method should not be used as the sole criterion for screening but will certainly indicate potential opportunities or problems within a given reservoir.

The second method, project economics, requires generation of an incremental oil recovery versus time estimate in order to evaluate the economic potential of a project. Due to the preliminary nature of the evaluation and the quality/quantity of data usually available at this stage, the economic analysis must include the effects of uncertainty so that upside potential and downside risk can be estimated. A combined engineering and economic evaluation of a reservoir using both these methods will give a valid screening of a prospect subject to the uncertainties of available data.

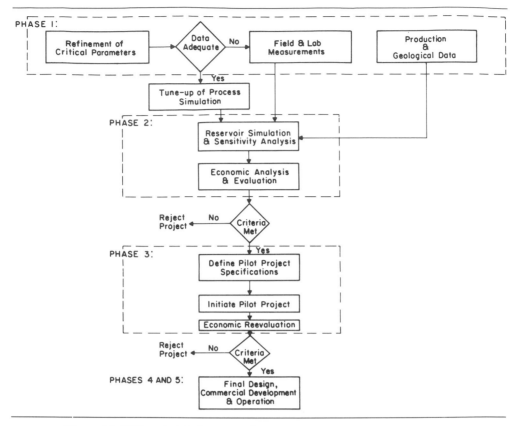

Figure 5.1 EOR project design approach.

5.1   **KEY RESERVOIR PARAMETERS.** A prospect is usually first screened by developing, within the available data, a comparison of key rock, reservoir, and fluid characteristics in technically and economically successful CO_2 floods with those in the candidate reservoir. While this preliminary screening is not a hard and fast rule in predicting the economic viability of an individual CO_2 flood, it may be indicative of a reduced or enhanced chance for success; that is, if a reservoir's properties do not fall in line with any previously successful application of CO_2 injection, forecast of economic implementation may be doubtful. Obversely, a potentially successful flood may not necessarily fulfill all reservoir parameter limitations.

Binary screening guides are inherently outdated from the onset. The number of field tests that are available is so small that a significant amount of new information is gathered from each project. Coupled with this dynamic addition of new technology are fluctuating economic factors that affect any decision-making process. However, because of their easy application, binary screening guides have become a popular preliminary screening tool. It is with this thought in mind that tables 5.1 and 5.2 are presented.

Table 5.1 Chronology of Miscible CO_2 Binary Screening Guides

Reservoir Parameter	Study							
	Geffen (1973)	Lewin and Associates (1976)	NPC (1976)	McRee (1977)	Iyoho (1978)	OTA (1978)	Carcoana (1982)	Taber and Martin (1983)
Viscosity, cp at reservoir conditions	<3	<12	≤10	<5	<10	≤12	<2	<15
Gravity, °API	>30	>30	≥27	>35	30 to 45	≥27[a] 27 to 30 ≥30	>40	>26
Fraction of oil remaining in area to be flooded (before EOR)	>0.25	>0.25		>0.25	>0.25		>0.30	>0.30
Depth, ft		>3,000	>2,300	>2,000	>2,500	≥7,200 ≥5,500 ≥2,500	<9,800[b]	>2,000
Temperature, °F		NC[c]	<250				<195	NC
Original reservoir pressure, psia	>1,100	>1,500					>1,200	
Permeability, md		NC		>5	>10		>1	NC

[a] See depth constraint.
[b] Due to temperature constraint.
[c] NC = not a critical factor.

Table 5.2 Criteria for the Application of CO$_2$ Enhanced Oil Recovery Methods

Specific Screening Parameters	CO$_2$ Miscible (Lewin) (1976)	Immiscible CO$_2$ (Klins and Farouq Ali) (1980)
Viscosity, cp at reservoir conditions	<12	100 to 1,000
Gravity, °API	>30	10 to 25
Fraction of oil remaining in area to be flooded (before EOR), % PV	>25	>50
Oil concentration, bbl/ acre-ft	>300	>600
Porosity times oil saturation	>.04	>.08
Depth, feet	>3,000	>2,300
Temperature, °F	NC[a]	NC
Original reservoir pressure, psia	>1,500	>1,000
Net pay thickness, ft	NC	NC
Permeability, md	NC	NC
Transmissibility (Permeability times thickness/viscosity)	NC	NC
GENERAL RESERVOIR PARAMETERS		
Thin pay preferred	Homogeneous formation preferred	
High dip preferred	No natural water drive	
Low vertical permeability in horizontal reservoirs	No major gas cap	
Natural CO$_2$ availability preferred	No major fractures	

[a]NC = not a critical factor.

Table 5.1 presents a chronology of screening guides for miscible applications. The characteristics of reservoirs into which CO$_2$ may be considered for miscible injection have changed very little over ten years. Light, low viscosity oils found in reservoirs over 2,000 to 3,000 ft deep are preferred. As discussed in chapter 1, this range of field parameters will make nitrogen and flue gas injection (higher injection pressures needed), hydrocarbon injection (less flexible oil type), and surfactant/polymer flooding (higher front-end cost and more design risk) competitive with miscible CO$_2$ injection in specific instances.

Table 5.2 compares the reservoir properties for miscible, light oil and immiscible, heavy oil reservoir parameters. This table, while not meant to serve as a permanent guideline, lists reservoir parameters that lie at or near the outer limit of those values considered economically feasible for immiscible and miscible CO$_2$ injection. The upper section of the table lists those parameters whose values can be placed in a fixed format. The lower section contains additional factors that cannot be quantitatively reviewed but applied on an individual field basis.

Note that these screening criteria are economic constraints rather than technological limitations. The Federal Energy Administration (FEA) reports that these parameters recognize that certain reservoir conditions tend to increase substantially the risk of economic failure of a project (Lewin and Associates 1976). These effects are due to the likelihood of low oil recovery relative to the investment and cost of injectant required. Examples of high economic risk parameters are reservoirs with active water drives or major gas caps, low residual oil saturations, low gravity, or extensive fractures.

It is again important to note that most reservoirs into which CO_2 could be injected successfully will not pass all the screening criteria. Each characteristic must be weighed and placed into the total picture. For example, a 15-cp oil should not be excluded automatically from CO_2 miscible displacement. Other factors may override this limitation. In properly designing a CO_2 injection project, it is important to remember that the total reservoir picture determines a project's success or failure. Therefore, the recovery prediction methods discussed in section 5.3, tied to an economic project evaluation, will give a more accurate assessment of a project's viability.

5.1.1   Carbon Dioxide Miscible Process

OIL VISCOSITY < 12 CP. As was shown in chapter 2, fractional flow and areal sweep are directly related to the oil viscosity. Since the fractional flow curve is shifted to the left with increasing oil viscosity, the flow of gas for a given gas saturation is significantly higher, thereby lengthening the timing of recovery and shortening gas breakthrough time, leading to higher amounts of produced gas to be handled. Also, the high mobility ratio inherent in gas displacements increases even further with an increase in oil viscosity, thereby reducing areal sweep efficiency.

The constraint of 12 cp is based on empirical observation. The vast majority of the current miscible CO_2 projects is operating in reservoirs with oil viscosities of 12 cp or less.

OIL GRAVITY > 30°. Reservoir oil with an API gravity of less than 30° is usually too viscous (as discussed) in relation to the injected CO_2. This leads to slower oil response and poorer sweep due to viscous fingering. Also, creation of a miscible bank in situ, be it by vaporization into a CO_2-rich vapor phase or extraction (condensation) of reservoir oil into a CO_2-rich liquid phase, requires reservoir crudes with a high percentage of intermediate hydrocarbons (C_5 to C_{20}), especially C_5 to C_{12}. Lower API gravity crudes usually have fewer of the components needed to form this miscible bank.

RESIDUAL OIL > 300 BBL/ACRE-FT. Determining oil saturation for a candidate reservoir is perhaps the most critical task in predicting economic viability. Residual oil saturation may be insufficient for oil banking (technological failure) or not banked in sufficient quantities for economic success. An in-depth study of this subject has been presented by the Interstate Oil Compact Commission in the text *Determination of Residual Oil Saturation* (1978). The commission describes the basic tools to predict residual oil saturation accurately, with a summary of their work shown in table 5.3.

Table 5.3 Basic Tools and Techniques to Determine Residual Oil Saturation

Tool	Technique	Can Be Used When Hole Is Cased	Has Been Field Tested	Expected Accuracy
Reservoir Performance	Volumetric determination	Yes	Yes	Poor[a]
Conventional cores	Saturation measurements from fresh cores	Core must be cut while drilling holes.	Yes	Poor
	Lab flooding techniques, imbibition, centrifuge, and so forth	Core must be cut while drilling holes.	Yes	Poor to fair
Pressure cores	Core with specially designed mud	Core must be cut while drilling holes.	Yes	Good to excellent
Backflow tracer	Backflow hydrolized tracers	Yes	Yes	Fair to good
Logging tools				
Resistivity	Conventional	No	Yes	Poor
	Log-inject-log	No	No	Good to excellent
Pulsed neutron	Conventional	Yes	Yes	Poor
	Log-inject-log with waterflood	Yes	Yes	Good to excellent
	Log-inject-log with chemical strip	Yes	Partially	Fair to good
Nuclear magnetism	Inject-log	No	Yes	Excellent
Carbon/oxygen	Conventional	Yes	Yes	Poor
Gamma radiation	Log-inject-log	Yes	No	Unknown (but could be excellent)
Dielectric constant	Conventional	No	Partially	Unknown

Source: Interstate Oil Compact Commission, *Determination* of Residual Oil Saturation, IOCC, Oklahoma City (1978).
[a]Expected accuracy (one standard deviation fractional pore volume) for rocks that have porosities greater than 0.25 fractional volume: excellent = less than 0.02, good = 0.02 to 0.04, fair = 0.04 to 0.06, and poor = greater than 0.06.

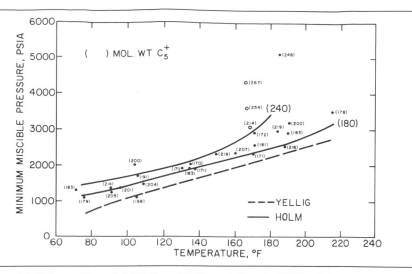

Figure 5.2 Holm and Josendal (1982) and Yellig and Metcalfe (1980) minimum miscibility pressure correlations for carbon dioxide ©SPE-AIME.

It is important to note that miscible CO_2 flooding may be either a secondary or a tertiary recovery process. Through simulation studies, Warner (1977) showed that significant amounts of oil can be recovered from previously waterflooded sands.

DEPTH > 3,000 FT. Miscibility is a direct function of reservoir pressure and temperature. This fact was discussed at length in chapter 3. If a minimum pressure of 1,500 psi is needed for miscibility and a general rule of thumb of 0.5 psi/ft of depth is accepted, then the minimum depth requirement is 3,000 ft.

PRESSURE > 1,500 PSI. Refer to the discussion on depth and to figure 5.2 by Holm and Josendal (1982) and Yellig and Metcalfe (1980). Certainly, as this figure points out, there may be specific instances where less than 1,500 psi is necessary for miscible displacement. However, 1,500 psi (3,000 ft) is perhaps a prudent design minimum to assure miscibility away from the wellbore where pressures are reduced from the sand-face injection pressure.

5.1.2   **CO_2 Immiscible Process**

OIL VISCOSITY 100 TO 1,000 CP. If CO_2 does not displace oil miscibly then it must rely on its ability to swell the crude oil and to sweep the reservoir more effectively than competitive processes such as water and polymer flooding. This increase in sweep is achieved by lowering the effective mobility ratio through a large reduction in oil vis-

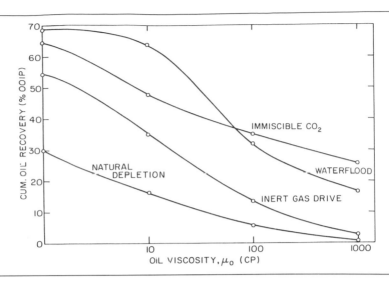

Figure 5.3 Oil recovery using four different processes and four different viscosity oils (Klins and Farouq Ali 1982).

cosity. By reducing the viscosity, not only is areal sweep increased but also the fractional flow curve to gas is shifted to the right, indicating a quicker oil response to CO_2 injection. As a result of work by Klins and Farouq Ali (1982) as shown in figure 5.3, immiscible CO_2 injection was clearly superior to waterflooding viscous crude of over 100 cp. The upper limit of 1,000 cp is purely arbitrary, suggesting that at some point, steam injection applications will override gains by CO_2 in thin pays.

OIL GRAVITY 10° TO 25° API. Reservoir oils with high API gravities usually have viscosities less than 100 cp. As was discussed earlier, reservoirs with lighter, less viscous crude oils may be best developed using alternate recovery schemes such as waterflooding if miscibility pressures cannot be attained, while heavier oil (<10° API) may be too viscous to displace economically with CO_2.

RESIDUAL OIL > 600 BBL/ACRE-FT (S_o > 0.5). From an economic standpoint, it is paramount that significant amounts of oil are in place prior to commencement of a CO_2 injection project. However, in viscous crude oil sands, if saturations are low, the injected gas may preferentially displace the water phase rather than banking the viscous oil. In terms of fractional flow displacements, initial saturations may be so far to the right and the fractional flow curve so steep that no tangent line—hence, no oil banking—will occur. As described in figure 5.4, Klins and Farouq Ali (1982) found this to be the case. For example, in the case of the 1,000-cp oil, recovery increased from 3% to 25% to 29%, as the initial oil saturation was increased from 40 to 60 to 70% respectively (compare recoveries of 55, 64, and 66% for the 1-cp oil).

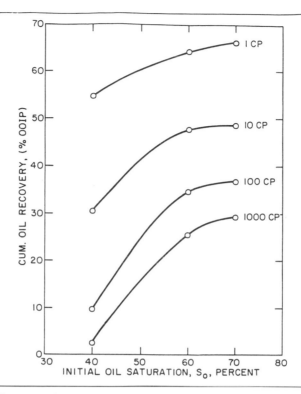

Figure 5.4 Oil recovery for four different viscosity oils as a function of initial oil saturation (Klins and Farouq Ali 1982).

DEPTH > 2,300 FT. It is doubtful that the pressures necessary for adquate swelling and viscosity reduction can be reached in shallower reservoirs. That is not to say that shallower reservoirs cannot be flooded with CO_2; however, at higher depths and pressures, economic displacement has greater potential since viscosity reduction and oil swelling are strong functions of displacement pressure.

PRESSURE > 1,000 PSI. See the preceding discussion concerning the depth constraint.

5.1.3   **General Screening Criteria (Miscible and Immiscible)**

THIN PAY OR HIGH DIP. If the injected CO_2 is lighter than the in-place oil or if CO_2 is injected with H_2O, gravity override can have an adverse effect on recovery efficiency. This process of gravity bypassing of reservoir oil is shown schematically in figure 5.5. Note that in the immiscible case, gravity override is severe. However, in the miscible case, mixing of fluids (CO_2-oil) causes a less drastic gravity difference between fluids

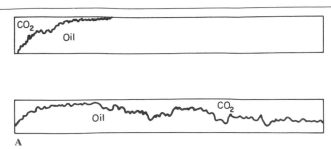

A

Figure 5.5A Sketch of secondary immiscible CO_2 displacement (Wang 1982).

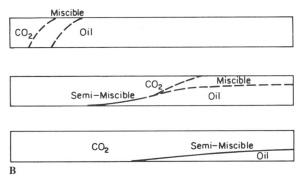

B

Figure 5.5B Sketch of secondary miscible CO_2 displacement (Wang 1982).

and, hence, a stabler flood front. Thin reservoirs (less than 30 ft) or steeply dipping beds are technically superior since a stable displacement front is more likely to develop. Early work in this area by Craig et al. (1957), Dumoré (1964) and Hawthorne (1960) was presented in chapter 2. More recent work by Wang (1982) using CO_2 as the displacing medium, is given in table 5.4. In downdip miscible CO_2 displacements, he realized an over 20 percentile increase in oil recovery as the dip angle increased from 0 to 30°.

Thin sands may also be advantageous in the case of immiscible viscous oil displacement since, due to excessive heat losses, thermal methods are not usually applicable in thin beds (less than 15 to 20 ft). However, there is one major drawback to flooding thin reservoirs, and that is the low STB/acre of initial oil in place, making the oil recovery target smaller.

LOW VERTICAL PERMEABILITY IN HORIZONTAL RESERVOIRS. Vertical permeability variation allows early breakthrough of gas in production wells from high permeability thief layers. Tighter sands will eventually be flooded out, but in fact, free gas production causes termination of a given CO_2 flood due to high gas-oil separation costs, thereby not allowing oil in lower permeability regions of the reservoir to be displaced. This

**Table 5.4 Downdip CO$_2$ Displacement of
Slaughter Estate Crude Oil at Temperature of 120°F**

Run Number	Pressure, psi	Dipping Angle, Degrees	Oil Recovery, Percent OOIP
SE-9	1,000	0	60.58
SE-10	1,000	15	72.72
SE-11	1,000	30	79.99
SE-1	1,500	0	66.83
SE-12	1,500	15	80.78
SE-13	1,500	30	89.78
SE-2	2,000	0	69.18
SE-14	2,000	15	80.31
SE-15	2,000	30	91.99

Source: G. C. Wang, "Microscopic Investigation of CO$_2$ Flooding Process," *J. Pet. Tech.* (Aug. 1982).

effect on vertical sweep efficiency was first quantified by Dykstra and Parsons (1950) and is designated by their coefficient of variation, V_{DP}. V_{DP} represents the amount of stratification in the reservoir where a value of 0 is a homogeneous reservoir. Their work, discussed in chapter 2, showed that breakthrough oil recoveries during waterflooding reduced drastically as permeability variation increases. A portion of their work, shown in figure 5.6, illustrates, for example, that at a mobility ratio of 10, sweep at water breakthrough can be halved as V_{DP} goes from 0 to 0.2. High vertical permeability also promotes gravity override. See the previous discussion on high dip or thin pay.

NATURAL CO$_2$ AVAILABILITY. One of the critical factors in CO$_2$ flooding projects, after preliminary field screening points toward a potential CO$_2$ application, is the CO$_2$ supply. Widespread flooding of reservoirs in the United States conceivably could require 40 to 50 trillion ft^3 of CO$_2$. Individual commercial-size projects may require well over 100 BSCF each. Potential sources of this gas are considered to be:

1. power plant stack gas,
2. by-products from ammonia or other chemical plants,
3. manufactured CO$_2$,
4. naturally occurring sources either from oil field acid gas separation or natural CO$_2$ deposits, and
5. by-products from future coal gasification plants.

Manufactured CO$_2$ or separation in one of these processes and its subsequent compression and transportation to the field appears to be prohibitively expensive. The most viable source of CO$_2$ for miscible or immiscible flooding appears to be from either existing known sources of naturally occurring CO$_2$ or from future discoveries. Several areas of known deposits are shown in figure 5.7 and are tabulated in table 5.5 from an extensive study of CO$_2$ resources and demand by Pullman-Kellogg (1978). Note that the largest deposits of naturally occurring CO$_2$ are found in the Four Corners area of Utah, Colorado, Arizona, and New Mexico.

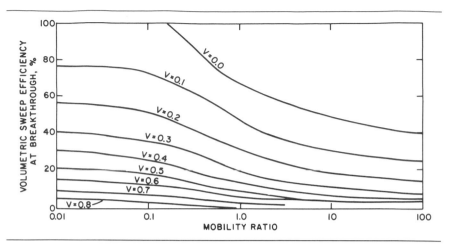

Figure 5.6 Volumetric sweep efficiency at breakthrough for a five-spot pattern with no initial gas saturation (Dykstra and Parsons 1950).

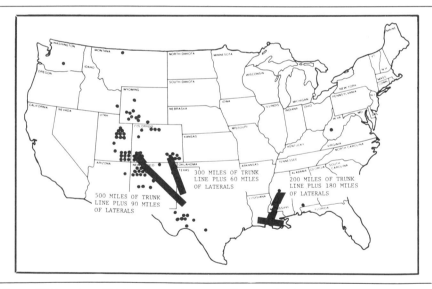

Figure 5.7 Location of high concentration naturally occurring CO_2 in the United States (Pullman-Kellogg 1978).

Also, since the Pullman-Kellogg (1978) study, drilling in southwestern Wyoming has indicated a large accumulation (>20 TSCF) of low Btu gas containing 60+ mole percent CO_2. Additional pipeline capacity to that shown on figure 5.7 is already under construction.

In addition to CO_2 availability, gas purity is an important variable. Contamination by other gases including nitrogen and methane were shown in chapter 3 to increase misci-

Table 5.5 Summary of Natural CO₂ Reserves

State	Reserves (BSCF CO₂)
Wyoming	3,900
Northern Utah	1,600
Northern Colorado	100
Southern Utah	1,000
Southern Colorado	3,900 to 5,900
New Mexico	7,000 to 10,600
Western Texas	600+
Eastern Texas	None
Louisiana	—
Mississippi	2,000 to 4,200
West Virginia	330
Los Angeles	None
Total	18,630 to 26,430
Selected Areas	
Jackson Dome, Miss.	3,000
Bravo Dome, N. Mex.	8,000 to 11,000
Sheep Mountain, Colo.	1,000 to 1,500
McElmo Dome, Colo.	3,000 to 5,000
Church Buttes, Wyo. (83% pure)	7,000
Gordon Creek, Utah	1,500 to 2,200
Farnham Dome, Utah	
Paradox Basin, Utah	1,000

Source: Pullman-Kellogg, Inc., "Sources and Delivery of Carbon Dioxide for Enhanced Oil Recovery," U.S. DOE Report FE-2515-24, Washington, D.C. (1978).

bility pressure and to create additional safety and production problems including having an adverse effect on compression and pipeline hydraulics due to the low compressibility of nitrogen and methane when compared to CO_2. Meanwhile, the addition of hydrogen sulfide and/or LPG decreases the minimum miscibility pressure but not without a significant increase in the safety and handling aspects of the CO_2-contaminant mixture.

Regardless of the additional safety/handling requirements, one project, the Slaughter Estate Unit tertiary pilot, used such a sour gas/CO_2 mixture. Because of difficulties in obtaining a reliable source of pure CO_2 for injection into the pilot, a solvent source (containing 72% CO_2 and 28% H_2S) from the nearby Slaughter gasoline plant was chosen. The necessary permits for high H_2S-content gas handling were obtained, and injection commenced in August 1976 (Rowe et al. 1983).

HOMOGENEOUS FORMATION. Unlike thermal recovery methods where the injected fluid need not contact the reservoir oil to displace it (the heat from the fluid does the actual displacing), miscible methods rely on fluid-oil contact for efficient displacement. Heterogeneities in the reservoir such as lenses, permeability pinchouts, etc. reduce the contacted volume and hence, productivity. This is true for all nonthermal injection processes.

NO NATURAL WATER DRIVE. An active, natural water drive usually leads to a relatively high recovery of the original oil in place in the most easily swept zones. This usually leaves considerably less incremental oil than needed for CO_2 displacement. Strong water drive may also inhibit the ability of CO_2 to reach, form, and drive an oil bank in the most potentially productive portions of the reservoir by carrying injected fluids away from those regions. High water saturations may or may not also trap oil in place by not allowing the CO_2 to come in contact with the remaining oil. This trapping process for miscible displacements in general was first described by Shelton and Schneider (1975), Raimondi and Torcaso (1964) and Stalkup (1970) and was discussed in chapter 2. However, Tiffin and Yellig (1982) and Stalkup (1983) have shown that the trapped oil effect may not be significant in CO_2 displacements where water is not injected with the CO_2.

NO MAJOR GAS SATURATION. In low dipping reservoirs, an initial gas cap may act as a thief zone for injected CO_2, allowing the injected fluids to override or bypass the residual oil. For steeply dipping reservoirs, it may be possible to inject CO_2 at the gas-oil contact and take advantage of gravity to stabilize the process. This recovery procedure would be similar to attic oil recovery by gas injection in Gulf Coast Salt Dome reservoirs.

A well-distributed gas saturation (i.e., no gravity segregation) will increase CO_2 injection rates, aid CO_2 movement, and accelerate oil recovery. However, this natural gas may cause poor oil banking and early breakthrough and will also tend to dilute the CO_2 in situ, resulting in a possible deleterious effect on phase behavior, including higher miscibility pressures.

NO MAJOR FRACTURES. Reservoirs with extensive fractures have a tendency to channel the injected fluid away from the oil-bearing portion of the reservoir. This drastically reduces the effectiveness of CO_2 drive and reduces the amount of oil recovered. However, with tracer studies and pressure-pulse testing, fracture orientation may be determined with patterns selected and oriented to take advantage of the fractures.

This does not eliminate the efficacy of small fracture stimulation treatments to increase drawdown and productivity in producing wells. Natural fractures may be advantageous in heavy oil, single well, huff-n-puff applications. Fractures can serve as paths to allow increased contact area between CO_2 and the reservoir oil. (CO_2 moves out through the fractures, then enters the rock matrix by diffusion and/or convection, reduces viscosity, introduces solution gas drive, and allows oil to drain from matrix to fractures.)

5.1.4 CO_2 Single Well Stimulation. As mentioned, CO_2 is also being considered as a single well stimulation rather than as a pattern drive fluid for heavy oil reservoirs. This process is not unlike cyclic steam injection (huff-n-puff). A slug of CO_2 is injected and oil viscosity is decreased by CO_2 solubility effects. CO_2 and oil are then produced back into the injection well. The cycle is then repeated as oil production tapers off.

Few data, to date, would indicate a dominance of CO_2 drive over single well stimulation or vice versa in heavy oil applications. However, preliminary work by Sankur and Emanuel (1983) indicates the drive process may be more effective in recovering 3,000-cp heavy oils, and the only full-scale project for the immiscible CO_2 process has been by CO_2 drive (Reid and Robinson 1981). Their study used core floods and computer simulations to investigate CO_2 drive and huff-n-puff techniques for tertiary recovery of a 14° API California crude. At 1,250 psia, CO_2 drive recovered up to 54% of the remaining oil in place at an efficiency ranging from 1.1 MSCF/bbl to 8 MSCF/bbl, depending on slug size.

The CO_2 huff-n-puff technique was tested on the same system at 1,250-psia pressure. The first cycle recovered 6% of the remaining oil in place, with a CO_2 utilization factor of 11 MSCF/bbl. The second cycle recovered an additional 3% with an efficiency of 33 MSCF/bbl.

A parametric black oil simulation study by Patton et al. (1982) predicted incremental oil recovery as a function of several operating and reservoir variables for the huff-n-puff scheme. Proper control of key *operating parameters* can enhance the potential profitability of the process. The most important are

CO_2 injected per cycle: A typical treatment will make use of between 0.1 and 0.2 MMSCF/ft of sand.

Number of cycles: CO_2's effectiveness decreases with the number of cycles. Patton et al. suggest that two to three cycles appear to be optimum.

Back pressure during production: Holding higher back pressures during the production cycle would suggest a potential for higher oil production exists due to greater solubility of CO_2 in the resident crude. However, Patton et al. noted just the reverse. For all oils studied, they observed a productivity increase with declining bottomhole pressure.

Treatment pressure: High treatment pressure forces more CO_2 into solution with a subsequent lowering of oil viscosity. They recommend that the well be treated at the highest feasible pressure.

Reservoir parameters dictate the selection of potentially commercial applications. The dominant factors are

Viscosity of the oil: Oil viscosities less than 2,000 cp are usually required for commercial application.

Critical (trapped) gas saturation: CO_2 will not flow in the reservoir until a critical gas saturation is built up. Therefore, as S_{gc} increases, more gas must be injected so the CO_2 will propagate deeper into the reservoir. Also, this immobile, residual gas phase reduces the relative permeability to oil.

However, these negative effects may be more than offset by the driving force exerted by an expanding CO_2 residual phase. Patton et al. (1982) conclude that the larger the treatment, the more beneficial the trapped gas phase becomes.

Oil saturation: Surprisingly, high oil saturations tend to reduce the efficiency of the process. This is due to the decision to evaluate the process for incremental production only. Gross oil recovery, however, is greatest for reservoirs having high oil saturations.

Water saturation: Patton et al. observed that incremental production is not adversely affected by high water saturations. Hence, the process is well suited to high water-cut reservoirs.

Permeability: Permeability shows a mixed effect on process efficacy. For oil viscosities greater than 1,000 cp, high reservoir permeability serves to enhance CO_2 stimulation. For lower oil viscosities, higher permeability lowers the process efficiency since more oil is produced under primary conditions.

However, as in the case of oil saturation, the high permeability reservoirs consistently produce more total oil (primary and stimulated) than their low permeability counterparts.

Wettability: A shift toward oil wetness, characterized by high water and low oil permeabilities for the same saturation, tends to reduce the effectiveness of the treatment.

Regression of Patton et al.'s (1982) results yielded the following equation of process efficacy (STB/MSCF):

$$E = 0.33 - 0.035N_c - 4.5 \cdot 10^{-5}\mu_o + 1.6 \cdot 10^{-4}P_t + 1.3 \cdot 10^{-9}P_t^2$$
$$+ 4.3 \cdot 10^{-5}k - 0.013S_{oi} - 0.69V_c, \tag{5.1}$$

where

E = efficacy of CO_2 cyclic stimulation process, STB incremental oil/MSCF CO_2 injected (0.01 to 0.97), (these are the variable ranges included in the correlation),

k = reservoir permeability, md (176 to 800),

N_c = number of cycles (1 to 5),

P_t = treatment pressure, maximum CO_2 bottomhole injection pressure, psia (350 to 1,800),

S_{oi} = initial reservoir oil saturation, fraction (0.59 to 0.75),

V_c = volume CO_2 injected per cycle per foot of sand, MMSCF/ft (0.05 to 0.75),

μ_o = oil viscosity, cp (177 to 28,000).

Although the statistical fit is good, use of this equation as well as the other preliminary screening criteria is recommended only as a guide in selecting candidate reservoirs. It should not be used in place of competent reservoir simulation to predict performance or history match field data in later, more inclusive studies.

5.2 **OBJECTIVES OF A PREDICTION MODEL.** The second, more comprehensive method of prospect screening involves a prediction of incremental recovery (rate) versus time and CO_2 breakthrough (rate versus time) combined with an evaluation of project economics. The objective is to provide meaningful estimates of after-tax present value profit (discounted cash flow, DCF) and rate of return (ROR) for the project. Due to the many uncertainties in reservoir data, recovery predictions, oil and gas prices, investment and operating costs, escalation, taxes, and monetary discount rates, the evaluation of project

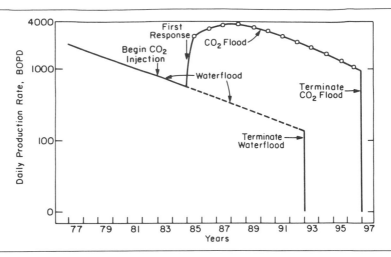

Figure 5.8 Comparison of production profiles for a CO_2 miscible flood versus continuation of an existing waterflood.

profitability must include the effect of uncertainty. In addition to mean value DCF and ROR, one need determine the project confidence interval, better known as upside potential and downside risk.

Figure 5.8 shows a typical incremental rate versus time for a miscible CO_2 flooding project and continuation of the current waterflood. The incremental investment and operating cost of the proposed CO_2 project must generate enough profit due to incremental production to justify early termination of the waterflood. In tertiary cases, where the waterflood is nearly complete, the cost to purchase and recycle CO_2 versus incremental recovery becomes the dominant economic consideration. Figure 5.9 illustrates that uncertainty of varying degrees are present in all parameters used to make the prediction of present value profit. However, the combination of all uncertainties can be translated into a probability distribution of present value (Monte Carlo simulation). The expected value at 50% probability is the mean DCF, and values at 10% and 90% probability represent an 80% confidence interval for the DCF.

The remaining sections of this chapter discuss hand-held methods (as opposed to numerical simulation) of generating oil recovery versus time curves for CO_2 floods. Chapter 6 will then discuss in more detail the economics and final design aspects of CO_2 flooding.

5.3   **RECOVERY PREDICTION METHODS.** The objectives of the recovery prediction methods are to compute:

1. oil production rate versus time (N_p versus V_{pi}),
2. CO_2 breakthrough versus V_{pi},

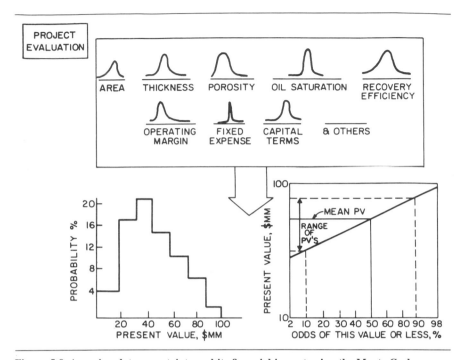

Figure 5.9 Assessing data uncertainty and its financial impact using the Monte Carlo Approach.

3. CO_2 production rate versus time (f_s versus V_{pi}), and
4. target oil ($E_M \cdot V_p$).

The recovery formulation was discussed in chapter 2 and takes the form

$$E_R = (E_A \cdot E_V \cdot E_D \cdot E_M)E_C \tag{5.2}$$

or

$$N_p = (E_A \cdot E_V \cdot E_D \cdot E_S)E_C \cdot V_p, \tag{5.3}$$

where

E_R = fractional recovery efficiency (oil produced/initial oil in place),
E_C = capture efficiency (oil produced/oil displaced),
E_A = areal sweep efficiency,
E_V = vertical sweep efficiency,
E_D = displacement efficiency,
E_M = mobilization efficiency,

and

$$E_S = \frac{S_{oi}}{B_{oi}} - \frac{S_{orp}}{B_{of}} = \left(\frac{S_{oi}}{B_{oi}}\right)E_M,$$ (5.4)

V_p = pattern pore volume, bbl,
N_p = cumulative oil produced, STB.

Research on the formulation of CO_2 recovery prediction methods is continuing at this time. This section presents the methods that have been investigated to date:

Lewin (1981) method:
    Tertiary CO_2 slug (0.7 HCPV),
    CO_2 recycle after breakthrough at 0.2 HCPV,
    Displacement efficiency based on match of field data;

Modified Koval (1963) method:
    Secondary ($S_{wi} = S_{wc}$),
    Line drive ($E_A = 1$),
    Five spot [E_A by Claridge (1972)],
    Trapped oil effect,
    Gravity stabilization,
    Continuous or slug injection;

Oil bank fractional flow (Stalkup 1983):
    Tertiary ($S_{oi} = S_{orw}$),
    Continuous CO_2 injection,
    Requires oil water relative permeabilities,
    Koval fingering of CO_2 through the oil bank;

General DOE/NPC CO_2 predictive model (Paul 1983):
    Secondary or tertiary,
    Continuous or WAG CO_2 injection,
    Koval mixing,
    Layer heterogeneity,
    Gravity override,
    Five-spot pattern only,
    No free gas.

5.3.1   **Lewin Method.** This tertiary method was published by DOE (Lewin and Associates 1981) and is the result of an empirical match of field recovery for a limited number of projects. The report does not describe the fundamentals used to develop the method, but it is claimed to be reasonably accurate. The method is restricted to tertiary projects based on 0.7 HCPV CO_2 slug injection followed by water. This, of course, is a serious limitation.

The following gives the recovery algorithm reformulated into the nomenclature of this discussion. The capture efficiency is set at 1 for waterflood sweep efficiencies of less than 0.6, indicating that CO_2 will follow the water path. For higher waterflood sweeps, the CO_2 is said to be not as efficient in contacting residual oil and will be penalized by setting $E_C = 0.75$.

The displacement efficiency, E_D, is an exponential function of the pore volumes injected with two empirical constants [eq. (5.7)]. The value of 0.2 HCPV prevents oil production prior to this volume of CO_2 injection and, therefore, represents breakthrough of the tertiary oil bank. The mobilization efficiency is similar to that discussed earlier with the restriction that Lewin uses $S_{CO_2} = 0.9(S_{orw})$ to determine the observed incremental recovery due to CO_2. This point is debatable, but it is fundamental to the empirical match.

$$N_p = E_C(E_A E_V E_D E_S)V_p, \tag{5.5}$$

where

$$
\begin{aligned}
E_C &= 1 \text{ for } E_A E_V < 0.6 \\
&= 0.75 \text{ for } E_A E_V \geq 0.6,
\end{aligned} \tag{5.6}
$$

$E_A E_V =$ waterflood sweep efficiency,

$$
\begin{aligned}
E_D &= 1 - exp^{[-5.4(V_{CO_2} - 0.2)]} \quad \text{if } (V_{CO_2} > 0.2) \\
&= 0 \quad \text{if } (V_{CO_2} \leq 0.2),
\end{aligned} \tag{5.7}
$$

$V_{CO_2} =$ cumulative injected volume of CO_2 (HCPV)

(see rate schedule on table 5.5),

$\text{HCPV} = (1 - S_{wc})V_p,$

$$E_S = \frac{S_{orw}}{B_{oi}} - \frac{S_{CO_2}}{B_{of}}, \tag{5.8}$$

$S_{orw} =$ waterflood residual oil saturation,

$S_{CO_2} =$ CO_2 flood residual oil saturation

$\approx 0.9 (S_{orw}),$

$B_{oi} =$ formation volume factor after H_2O flood,

$B_{of} =$ formation volume factor after CO_2 flood,

$V_p =$ pattern pore volume, bbl

$= 7,758 \, \phi \cdot H \cdot A,$

$N_p =$ cumulative recovery, STB. $\tag{5.9}$

The method also assumed a typical project schedule as shown in table 5.6. CO_2 is assumed to break through after 0.2 HCPV is injected and to be totally recycled after 0.3 HCPV is injected. The net purchase of CO_2 is 0.25 HCPV with a total slug injection of 0.7 HCPV followed by water. The HCPV used here is defined as $(1 - S_{wc})V_p$.

Table 5.7 shows a sample calculation for North Cowden based on the following reservoir properties:

Pattern area   40 acres
Net thickness   60 ft

203

Table 5.6 Lewin (1981) Model: Purchase/Recycle Schedule

Years	HCPV CO$_2$ Purchased	HCPV CO$_2$ Recycled	Cumulative HCPV Injected
1	0	—	0
2	0.10	—	0.1
3	0.10	—	0.2
4	0.05	0.05	0.3
5	—	0.10	0.4
6	—	0.10	0.5
7	—	0.10	0.6
8	—	0.10	0.7
9	—	—	0.8
10	—	—	0.9
11	—	—	1
12	—	—	1.1

Note: For economic purposes, year 1 is considered project set-up only, no production or injection.

Table 5.7 North Cowden Tertiary Recovery Prediction: Lewin (1981) Method

Year	Cumulative Injection, V_{CO_2}	Displacement Efficiency, E_D	Cumulative Oil Recovery N_P, MSTB	Yearly Oil Recovery, ΔN_P, MSTB
1	0	0	0	0
2	0.1	0	0	0
3	0.2	0	0	0
4	0.3	0.417	43.6	43.6
5	0.4	0.660	69	25.4
6	0.5	0.802	83.9	14.9
7	0.6	0.885	92.5	8.7
8	0.7	0.933	97.6	5
9	0.8	0.961	100.5	2.9
10	0.9	0.977	102.2	1.7
11	1	0.987	103.2	1
12	1.1	0.992	103.8	0.6
Total				103.8

Maximum recovery

$E_C = 1,$

$E_A E_V = 0.54,$

$E_S = \dfrac{0.350}{1.08} - \dfrac{0.315}{1.62} = 0.13,$

$V_P = 7758(0.08)(60)(40) = 1490$ Mbbl,

$N_P = 1(0.54)\ E_D(0.13)(1490)$

$\quad = 104.6 E_D$ (MSTB)

Note: For economic purposes, year 1 is considered project set up only, no production or injection.

Porosity	8%
Waterflood sweep ($E_A E_V$)	54%
S_{wc}	20%
S_{oi}	80%
S_{orw} (residual to H_2O flood)	35%
S_{orCO_2} (residual to CO_2 flood)	31.5%
B_{oi} (at start of CO_2 injection)	1.08 bbl/STB
B_{of} (at end of CO_2 injection)	1.62 bbl/STB
Oil viscosity	1.4 cp
CO_2 viscosity	0.2 cp

The maximum recovery ($E_D = 1$) is 104.6 MSTB based on the specified mobilization efficiency. Again, the first year shows no CO_2 injection and is used for capital investment in the economic analysis.

5.3.2   **Modified Koval Method.** This secondary recovery method is based on the original work of Koval (1963) as modified by Claridge (1972) for areal sweep and is modified further for trapped oil saturation and Hawthorne's (1960) gravity stabilization effect. The following summarizes Koval's work as discussed in chapter 2 for linear displacements (see also Paul 1983). The method was extended to include areal sweep by Claridge and is based upon the concept of apparent pore volume.

$$N_P = E_C[E_A(E_V E_D)(S_{oi} - S_{or})]V_p, \tag{5.10}$$

where

$$E_C = 1 \text{ (oil produced/oil displaced)},$$
$$E_A = \text{five-spot areal sweep (Claridge 1972)},$$
$$E_A = 1 \text{ (linear systems)},$$

$$(E_V E_D) = \frac{2(KV_{pvd})^{1/2} - 1 - V_{pvd}}{K - 1}, \tag{5.11}$$

$$K = HFE = HF\left[0.78 + 0.22\left(\frac{\mu_o}{\mu_s}\right)^{1/4}\right]^4, \tag{5.12}$$

$$\log_{10} H = [(V_{DP}/(1 - V_{DP})^{0.2})] \text{ (stratified reservoirs)}, \tag{5.13}$$
$$H = 1 \text{ (homogeneous reservoirs)},$$
$$F = 1 \text{ (no gravity override)}, \tag{5.14}$$
$$F = 0.565 \log_{10}(t_h/t_v) + 0.870 \text{ (such that } F \geq 1), \tag{5.15}$$
$$\frac{t_h}{t_v} = C\left(\frac{k_v}{k_h}\right)\left(\frac{A}{h}\right)(k_h h)\frac{\Delta\rho}{Q\mu},$$

where

$$k_v = \text{vertical permeability, md},$$

k_h = horizontal permeability, md,
A = pattern size, acre,
h = reservoir thickness, ft,
$\Delta\rho$ = water-CO_2 density difference, g/cc,
Q = injection rate, res bbl/d,
μ = CO_2 viscosity, cp,
C = pattern constant (2.5271 for five-spots or 2.1257 for line drives).

S_{oi} = initial oil saturation prior to CO_2 injection,
S_{or} = residual oil saturation after CO_2 injection,
V_p = pore volume of pattern, bbl,
V_{pi} = actual pore volumes of solvent injected (note: $V_{pi} = V_{pvd}$ for linear systems, $E_A = 1$),
V_{DP} = Dykstra-Parsons factor (see chapter 2),
V_{pvd} = invaded pore volumes injected.

Breakthrough:

$$V_{pvd} = \frac{1}{K}. \tag{5.16}$$

Solvent fractional flow to find V_{pvd} at abandonment:

$$f_S = \frac{K - (K/V_{pvd})^{1/2}}{K - 1} \text{ after breakthrough.} \tag{5.17}$$

Defining $E_{vd} = (E_V E_D)$ as given in equation (5.11), the apparent pore volume injected, V_{pa}, is related to the actual pore volume injected by, V_{pi},

$$V_{pa} = \frac{V_{pi}}{E_{vd}}. \tag{5.18}$$

This term is used to evaluate areal sweep by Claridge's correlation that has the form

$$E_A = \frac{(U_1 + 0.4 \, U_2)}{(1 + U_2)}, \tag{5.19}$$

where U_1 and U_2 (see section 5.5) are empirical functions of the mobility ratio (M) and the apparent pore volumes injected (V_{pa}).

If we divide the actual pore volume injected by the invaded area fraction, we obtain the pore volumes injected relative to the invaded area:

$$V_{pvd} = \frac{V_{pi}}{E_A} = \left(\frac{E_{vd}}{E_A}\right)V_{pa}, \tag{5.20}$$

which is used to evaluate E_{vd} in equation (5.11) for nonlinear displacements. The procedure converges in an iterative fashion. Section 5.5 presents the actual solution approach in more detail. Table 5.8 shows such an iterative calculation for secondary recovery using CO_2 in one inverted, 40-acre (confined) five spot of the North Cowden field. The reader is referred to section 5.3.1 for a more complete reservoir description.

For this case, assumptions have been made that the sand is homogeneous ($H = 1$), that there is no gravity override ($F = 1$), and that CO_2 is continuously injected. The reader is referred to Claridge (1972) for additional information concerning displacements of a single slug of CO_2 followed by continuous water injection for miscible secondary CO_2/oil methods.

In order to account for trapped oil and gravity stabilization, we returned to the fractional flow definitions discussed in chapter 2. Assume that the solvent fractional flow is given by

$$f_s = \frac{K S_s (1 + G)}{1 - S_{ot} + S_s(K - 1)}. \tag{5.21}$$

If the gravity term, G, is zero and the trapped oil saturation, S_{ot}, is zero, then this equation reduces to equation (2.8).

Since the derivative of f_s equals $1/V_{pi}$, we get

$$\frac{1}{V_{pi}} = \frac{K(1 + G)}{[1 - S_{ot} + S_s(K - 1)]^2}. \tag{5.22}$$

Solving for S_s and substituting in equation (5.21) yields the following relationships:

$$(E_V E_D) = \frac{2[(1 + G)KV_{pi}]^{1/2} (1 - S_{ot}) - 2 - (1 + KG)V_{pi} + (1 - S_{ot})^2}{K - 1}, \tag{5.23}$$

S_{ot} = trapped oil saturation,

G = gravity term (Hawthorne 1960) $\qquad\qquad$ (5.24)

$$= \frac{k_H k_{ro} A g}{\mu_o Q_t} (\rho_s - \rho_o) \frac{\sin \beta}{\cos (\alpha - \beta)},$$

where

α = formation dip angle,
β = interface dip angle,
Q_t = total throughput rate,
k_H = horizontal (dipped) permeability,
K = Koval factor = HE,
V_{pi} = pore volumes solvent injected,
ρ_s = solvent density,
ρ_o = oil density.

Breakthrough:

$$V_{pi} = \frac{(1 - S_{ot})^2}{(1 + G)K}. \tag{5.25}$$

Table 5.8 North Cowden Secondary Recovery Prediction: Modified Koval (1963) Model

Actual Pore Volumes Injected, V_{pi}	Apparent Pore Volumes Injected, V_{pa}	Areal Sweep Efficiency, E_A	Displacement Efficiency, $E_V E_D$	Cumulative Oil Recovery, (BBL)
0.296	0.48	0.475	0.623	214,009
0.344	0.50	0.493	0.688	245,171
0.414	0.55	0.527	0.753	287,161
0.481	0.60	0.558	0.802	323,338
0.547	0.65	0.584	0.842	355,535
0.612	0.70	0.608	0.875	384,249
0.676	0.75	0.629	0.902	409,888
0.739	0.80	0.648	0.924	432,853
0.800	0.85	0.667	0.941	453,508
0.860	0.90	0.683	0.956	472,082
0.920	0.95	0.698	0.968	488,439
0.980	1	0.706	0.980	499,775
1.036	1.05	0.720	0.986	513,477
1.069	1.08	0.728	0.990	520,689

Breakthrough recovery

$$K = [0.78 + 0.22(1.4/0.2)^{1/4}]^4$$
$$= 1.676,$$
$$(V_{pvd})_{BT} = 1/K = 0.597,$$
$$(E_V E_D)_{BT} = 1/K = 0.597,$$
$$(E_A)_{BT} = (1 + 0.4M)/(1 + M)$$
$$= 0.475,$$
$$(V_{pi})_{BT} = (0.475)(0.597)$$
$$= 0.284,$$
$$(V_{pa})_{BT} - (V_{pi})_{BT}/(E_V E_D)_{BT}$$
$$= 0.284/0.597 = 0.476.$$

These figures give us an approximate starting point for the iterative technique described in section 5.5.

Ultimate recovery

Assuming $(f_S)_{ABAND.} = 0.9$ and $E_C = 1$, from equation (5.20),

$$(V_{pvd})_{ABAND.} = 1.47,$$

$$(E_V E_D) = \frac{2[K(V_{pvd})_{ABAND.}]^{1/2} - 1 - V_{pvd}}{K - 1}$$

$$= .990.$$

At abandonment, truncate table 5.8 when $(E_V E_D) = 0.990$. Therefore,

$$(V_{pi})_{ABAND.} = 1.069,$$
$$(V_{pa})_{ABAND.} = 1.08,$$
$$(E_A)_{ABAND.} = 0.728,$$
$$N_P = 1[(0.728)(0.990)(0.8 - 0.315)](1,490,000)$$
$$= 520,700 \text{ bbl.}$$

Solvent fractional flow:

$$f_s = \frac{K(1 + G) - [(1 + G)K/V_{pi}]^{1/2}(1 - S_{ot})}{K - 1}. \tag{5.26}$$

Total recovery:

$$V_{pi} = \frac{(1 + G)K(1 - S_{ot})^2}{(KG + 1)^2}. \tag{5.27}$$

The trapped oil saturation, S_{ot}, may be estimated using a modified Raimondi-Torcaso (1964) equation such that

$$S_{ot} = S_{orw}[1 + \alpha (k_{ro}/k_{rw})], \tag{5.28}$$

where Claridge's (1982) wetting factor, α, has a value of 5 to 25 for partially oil-wet dolomites.

All of this development applies only to a case where water is immobile ($S_{wi} = S_{wc}$), which corresponds to a secondary flood. Extension to tertiary floods with implied oil banking requires relative permeability between oil and water, as discussed in the next section.

5.3.3   **Oil Bank Fractional Flow (Linear Systems).** One critical problem in developing a tertiary recovery predictive model is to compute the delay time prior to oil breakthrough. In the Lewin model, this was set at 0.2 HCPV, which may be reasonable for some field projects but not for others. Delay time is indicative of oil bank formation due to fractional flow of the miscible CO_2-oil mixture and water.

The formulation of this model assumes that a miscible CO_2/oil bank will displace water in an amount equal to the pore volumes of CO_2 injected prior to breakthrough. This corresponds to an incompressible system, which is obviously not the case in practice but should be sufficient to investigate the validity of the approach. The formulation also allows CO_2 to finger through the oil using the modified Koval approach. Figure 5.10 shows a schematic of the tertiary oil bank displacement process with fingering for continuous CO_2 injection. At this point, the mobile water is being flooded by the oil bank, and breakthrough is yet to occur.

The average water saturation at breakthrough, $\overline{S}_{wbt}$, can be computed based on tangent analysis similar to the Welge method for a waterflood (1952), but using a reverse of the fractional flow curve between oil and water. In the following equations are the details of the oil bank formulation.

Water fraction:

$$f_w = \frac{k_{rw}/\mu_w}{k_{rw}/\mu_w + k_{ro}/\mu_o}, \tag{5.29}$$

$$k_{rw} = k_{rw}^o \left(\frac{S_w - S_{wc}}{1 - S_{wc} - S_{orw}}\right)^{Nw}, \tag{5.30}$$

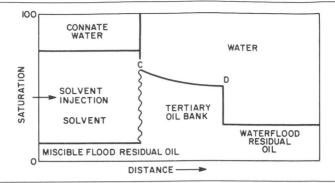

Figure 5.10 Idealized saturation profile for the linear displacement of waterflood residual oil by continuous solvent injection.

$$k_{ro} = k_{ro}^{\circ}\left(\frac{1 - S_w - S_{orw}}{1 - S_{wc} - S_{orw}}\right)^{N_o}.$$ (5.31)

where N_w and N_o are constants used to fit the relative permeability data. Also, k_{rw}° is the relative permeability to water at waterflooded residual oil saturation and k_{ro}° is the relative permeability to oil at the connate water saturation. Now

Derivative:

$$\frac{df_w}{dS_w} = \left(\frac{f_w}{k_{rw}}\right)^2\left(\frac{\mu_w}{\mu_o}\right)k_{ro}\left(\frac{dk_{rw}}{dS_w} - k_{rw}\frac{dk_{ro}}{dS_w}\right),$$ (5.32)

where

$$\frac{dk_{rw}}{dS_w} = \frac{N_w k_{rw}}{S_w - S_{wc}},$$ (5.33)

$$\frac{dk_{ro}}{dS_w} = \frac{-N_o k_{ro}}{1 - S_w - S_{orw}}.$$ (5.34)

Breakthrough:

$$\left(\frac{df_w}{dS_w}\right)_f = \frac{1 - f_{wf}}{(1 - S_{orw}) - S_{wf}},$$ (5.35)

$$\overline{S}_{wBT} = (1 - S_{orw}) - \frac{1}{\left(\dfrac{df_w}{dS_w}\right)_f},$$ (5.36)

$$V_{iBT} = S_{wi} - \overline{S}_{wBT},$$ (5.37)

$$f_w = 1 \text{ for } V_{pi} \leq V_{iBT}.$$ (5.38)

After breakthrough:

$$\left(\frac{df_w}{dS_w}\right)_2 = \frac{1}{V_{Pi}},$$

(5.39)

$$f_h = 1 - f_w(S_{w2})$$

(5.40)

$$= f_o + f_s,$$

(5.41)

$$\overline{S}_w = S_{w2} + (V_{pi})(f_H).$$

Modified Koval:
If $f_s > 0$ at breakthrough,

$$\alpha = KV_{pi},$$

(5.42)

$$V_p^* = V_{pi}/\alpha,$$

(5.43)

And f_s is evaluated based on pseudo–pore volume V_p^*. Prior to breakthrough, the fractional flow of water is 1. After breakthrough, a shock occurs, with the fractional flow of hydrocarbon phase (oil plus CO_2) dominating. The CO_2 is assumed not to finger through the flood front interface with a corrected pseudo–pore volume adjustment.

One disadvantage of this preliminary screening approach is that it requires relative permeability data, which may not always be available, particularly for early-on predictions.

5.3.4   **Oil Bank Fractional Flow (Five-Spot Systems).** Stalkup (1983) has presented an approach similar to the continuous tertiary CO_2 displacements discussed in section 5.3.3 with an extension for vertical and areal sweepout. The first step in the solution scheme is to determine graphically the average oil saturation in the oil bank and the water saturation at the trailing edge of the oil bank (points D and C on figures 5.10 and 5.11).

Water saturation at the front edge of the tertiary oil bank is determined by the point of tangency to the water/oil fractional flow curve of a line passing through the point $f_w = 1$, $S_w = 1 - S_{orw}$. This is shown as point D on figure 5.11. Water saturation at the trailing edge of the oil bank, S_{wt}, is represented by point C. This is determined by a point of tangency to the water/oil fractional flow curve from a line passing through point A ($f_w = 1$, and $S_w = 1 - S_{orm} + S_{orm}R_{so} + S_w^*R_{sw}$). S_w^* is the intersection of line AB and the water/solvent fractional flow curve (point E) and must be solved iteratively when $R_{sw} > 0$. The average saturation in the oil bank, S_{ob}, is estimated simply as one minus the arithmetic average of the leading and trailing edge water saturations.

To estimate volumetric sweepout as a function of the pore volumes injected, let us first review the notion of gravity stability and viscous fingering. Three flow regimes have been discussed in chapter 2 and are shown in figure 5.12. Region I and II type displacements are characterized by a single gravity tongue overriding the oil. When some stable velocity has been exceeded, a transition zone consisting of a gravity tongue and viscous fingers develops. Region IV type displacements, when flood velocity is higher than a critical value, are dominated by viscous fingering and exhibit no gravity

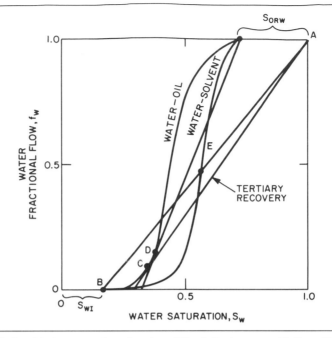

Figure 5.11 Graphical construction of tertiary CO$_2$-oil displacements (Stalkup 1983).

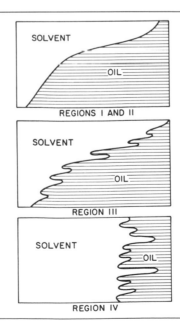

Figure 5.12 Flow regimes for miscible displacement in a vertical cross section (Stalkup 1983) ©SPE-AIME.

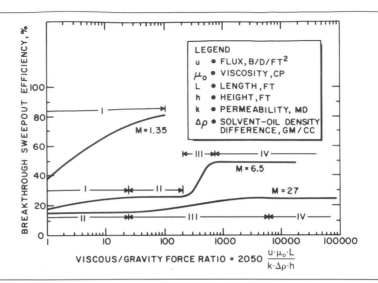

Figure 5.13 Flow regimes in a two-dimensional, uniform linear system and their relationship with sweepout efficiency (Stalkup 1983) ©SPE-AIME.

override. Figure 5.13 further illustrates the different regimes in *horizontal*, miscible displacements and shows how breakthrough sweepout is affected by flow regime type and the viscous/gravity force ratio. As the viscous/gravity force ratio is increased, the effect of gravity override is diminished while viscous fingering dominates and sweepout improves. Unless oil and CO_2 densities are similar, it may be difficult to operate miscible CO_2 displacements at a viscous/gravity ratio in excess of 100. This would lead one to assume that gravity override will be a dominant factor in horizontal CO_2 displacements if vertical communication exists.

For *down-dip*, gravity stable displacements the reader is referred to Dumoré's work cited in chapter 2. In this type of displacement, slower flood velocities allow gravity to diminish the override effect and hence vertical sweep is improved. This is just the opposite for horizontal floods where, in general, higher flood rates improve vertical sweep.

Stalkup (1983) has noted that when sweepout is dominated by tonguing, few experimental data are available. Figure 5.14 shows volumetric sweepout estimated by Stalkup for a horizontal five-spot pattern exhibiting Region II flow regime characteristics assuming the effect of transverse mixing is negligible.

With continuous CO_2 injection in tertiary recovery then, the following steps remain to estimate recovery. For a given displaceable volume of solvent injected, sweepout of the leading edge of the oil bank is estimated by calculating a pseudodisplaceable volume injected as follows:

$$D_{vob} = D_{vs} + \Delta D_{vob}, \tag{5.44}$$

$$\Delta D_{vob} = D_{vs}\frac{(S_{orw} - S_{orm})}{(S_{ob} - S_{orw})}, \tag{5.45}$$

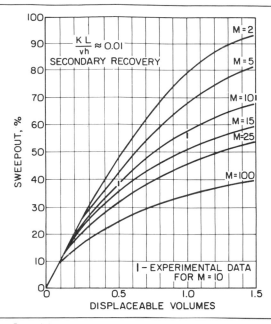

Figure 5.14 Volumetric sweepout in a normal five-spot pattern for Region II flow (Stalkup 1983) ©SPE-AIME.

and

$$D_{vs} = \frac{Q_s}{V_p(1 - S_{orm} - S_{wt})},$$ (5.46)

where

D_{vob} = pseudodisplaceable volumes for estimating oil bank sweepout,
D_{vs} = displaceable volumes of solvent injected,
S_{ob} = average oil saturation in the oil bank, fraction,
S_{wt} = water saturation at the trailing edge of the oil bank, fraction,
S_{orw} = waterflood residual oil saturation, fraction,
S_{orm} = miscible flood residual oil saturation at reservoir conditions, fraction,
V_p = pattern pore volume, RB,
Q_s = volume of solvent injected, RB.

Last, oil recovery is calculated from

$$N_p = V_p E_s(S_{orw} - S_{orm}) - V_p(E_{ob} - E_s)(S_{ob} - S_{orw})/B_o$$ (5.47)

where

N_p = cumulative recovery from the start of CO_2 injection, STB,

E_s = volumetric sweepout of solvent, fraction (use D_{vs} and figure 5.14),
E_{ob} = volumetric sweepout of leading edge of the oil bank, fraction (use D_{vob}),
B_o = oil formation volume factor at beginning of the miscible flood, RB/STB.

The reader is now referred to Stalkup (1983) for sample calculations.

5.3.5   **General DOE/NPC Carbon Dioxide Predictive Model.** A major drawback of the Lewin, Koval, and fractional flow models is their inability to handle the WAG injection process and/or to compute the delay time prior to oil breakthrough in tertiary floods. With this in mind, the U.S. DOE contracted the development of a new CO_2 miscible flood predictive model (CO2PM) (Paul 1983).

The model and a lengthy discussion of its development will not be presented here. Its assumptions, limitations, and applicability to handling the complex mechanistic calculations usually available only through full numerical simulation with the low operating cost of less technically sophisticated hand-held calculator models are discussed.

Paul (1983) writes,

The oil rate and recovery algorithm includes treatments for a number of effects and conditions: fractional flow, areal sweep, reservoir heterogeneity, viscous fingering, and gravity segregation. A one-dimensional fractional flow theory was developed for first-contact miscible displacements in the presence of a second immiscible phase. The theory is based on a specialized version of the method of characteristics known as *coherence* or *simple wave theory*. The theory also incorporates the Koval factor method to account for unstable miscible displacements (fingering). An extension of the Koval approach is used to model the influence of gravity segregation. Reservoir heterogeneity is accounted for by allowing up to five layers in the model. The fractional flow theory with gravity and heterogeneity dependence is corrected for areal sweep with a generalization of Claridge's procedure to first-contact miscible floods of arbitrary WAG ratios and arbitrary initial conditions.

The model assumes that the displacement is first-contact miscible and that the Koval factor method adequately describes fingering and gravity effects. Due to boundary condition restrictions in the method of characteristics, the CO2PM cannot be used with a Koval factor below about 2 (very stable miscible displacement). The model does not calculate fluid injectivity and assumes that the reservoir can take any specified injection rate. The model also assumes steady-state displacement with constant injection rate and WAG ratio.

The CO2PM can be used for a wide range of initial reservoir conditions. It can be used for both secondary (mobile oil present) and tertiary recovery. It can predict performance for continuous CO_2 injection or for any WAG process. Ultimate chase fluid is limited to water. The model is limited to five-spot patterns.

In the CO2PM, CO_2 and water are simultaneously injected in a proportion determined by the WAG ratio. This is not a true WAG process, but the overall recoveries should not be influenced by this factor. The reservoir fluid properties and formation volume factors are kept constant during the oil recovery calculations, but this assumption should not affect the results in the majority of cases. The fractional flow calculations are performed under the assumption of no free gas saturation in the reservoir. Finally, the model may not accurately predict the bypassing of heavy oils by more mobile CO_2. However, in such cases, the displacement most likely will be immiscible. [1983, pp. 2–3]

5.4   **MODEL VALIDATION.** Each of the recovery prediction methods described in the last section has been validated using at least one of the four possible methods:

1. Comparison to analytic solutions,
2. Comparison to laboratory data,
3. Comparison to numerical simulator results,
4. Comparison to field project results.

This section discusses the validation of each method.

5.4.1   **Lewin Method.** The Lewin (1981) method was developed by an empirical fit of recovery curves from field projects. No details of their validation procedures are currently available. The model duplicated oil recovery volumes for the Crossett, North Cowden, and Keystone tests. However, produced CO_2 volumes appear to be in error.

5.4.2   **Modified Koval Method.** The modified Koval (1963) method is strictly a secondary method and assumes that no mobile water ($S_{wi} = S_{wc}$) is present. The method described in the previous section has been verified by comparison with laboratory data.

Figure 5.15 shows the results of Blackwell et al. (1959) for *linear* miscible floods in sand pack models of various dimensions. Koval's K-factor correlation gives a reasonable match but cannot account for geometry effects by the nature of the method. Figure 5.16 shows a comparison with Blackwell's data for recovery at breakthrough. Since these are the same data originally used by Koval, they can be used as a check to verify that the user's calculations are correct.

Figure 5.17 shows a comparison with Lacey et al.'s (1961) data for miscible floods at viscosity ratios of 10, 41, and 85 in a glass bead *five-spot* model. The match achieved by the modified Koval method is shown. The sweep efficiency correlation by Claridge (1972) is subject to statistical error, and small adjustments could allow this method to achieve an excellent match. Alternately, other correlations for sweep efficiency could be tried.

5.4.3   **Oil Bank Fractional Flow (Linear Model).** The oil bank fractional flow tertiary method has been verified using the linear floods reported by Doscher and El-Arabi (1981). The initial oil saturation was about 26%, with connate water saturation estimated to be 24 to 27%. No relative permeability data were available, so an adjustment of the curvature and end points was made to achieve a reasonable match. The curves used for the match

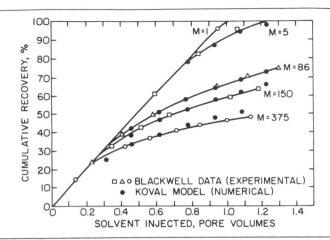

Figure 5.15 Effect of mobility ratio on cumulative oil recovery showing Blackwell et al.'s data for a line drive (1959).

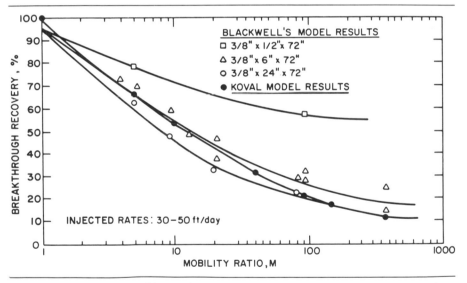

Figure 5.16 Blackwell et al.'s experimental data (line drive) on the effect of mobility ratio on oil recovery at breakthrough as compared to a Koval-type numerical prediction (1959).

are shown on figure 5.18. The resulting computation of oil, water, and CO_2 recoveries versus CO_2 injected, as shown on figure 5.19, shows a reasonable match with the data. The relative permeability curves used cause breakthrough to occur about 0.05 PV after the observed breakthrough, but the overall recoveries are in reasonable agreement with the data.

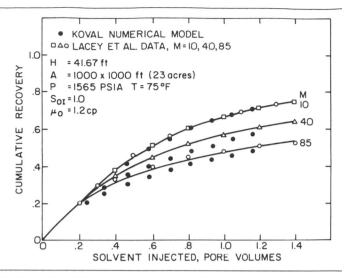

Figure 5.17 Effect of mobility ratio on cumulative oil recovery showing Lacey et al.'s experimental data for a five-spot as compared to Koval-type numerical results (1961).

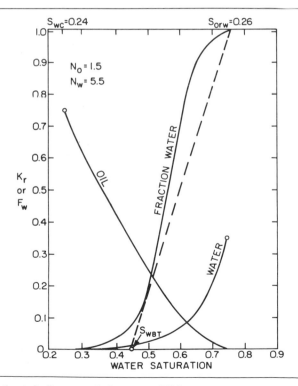

Figure 5.18 Estimated oil-water relative permeabilities.

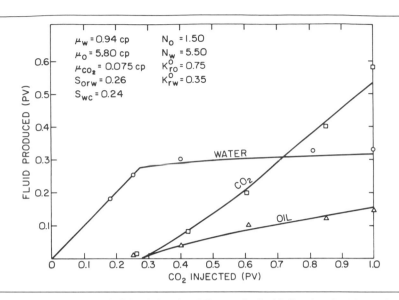

Figure 5.19 Comparison of oil-bank fractional flow method with Doscher-Arabi data for tertiary flood of a sandpack (1981).

5.4.4   **Oil Bank Fractional Flow (Five-Spot Systems).** Stalkup (1983) notes that few data exist against which the accuracy of his procedure can be tested. Figure 5.20 compares the method with results of a vertical cross-sectional calculation made with a finite-difference miscible flood simulator for Region II gravity-tongue-dominated flow. In Stalkup's example, first production of incremental oil was estimated by the simple method to occur nearly 0.1 HCPV of solvent injection earlier than calculated by the simulator. After about 0.3 HCPV of injection, oil recovery calculated by the two methods after a given hydrocarbon pore volume of injection was in fair agreement. Excessive numerical dispersion in the simulator may have moderated the effective solvent/oil viscosity ratio unduly and caused solvent sweepout in the numerical model to be too large.

In figure 5.21, Stalkup compares actual oil recovery from the Little Creek CO_2-flood pilot test with oil recovery estimated by this procedure. In this example, also, the desk-top method gave an optimistic prediction; that is, predicted recovery occurred sooner than in the pilot test and was higher than in the pilot test for a given hydrocarbon pore volume of solvent injected up to 1 HCPV.

5.4.5   **General Carbon Dioxide Predictive Model.** Paul (1983) presents a comparison of DOE/NPC CO_2 predictive model results with one field test and one numerical simulation.

The actual and predicted responses for the Slaughter Estate Unit Tertiary Pilot were compared in several runs. The best fit of observed data to predicted response is shown in figure 5.22. This match used three layers and proved far superior to a single-layer model.

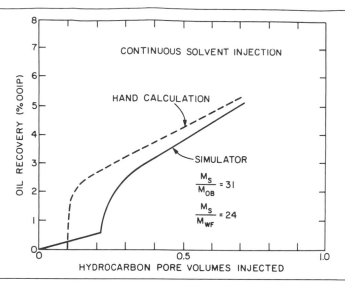

Figure 5.20 Comparison of desktop and finite-difference simulator calculations (Stalkup 1983) ©SPE-AIME.

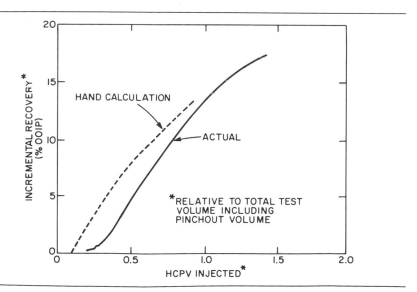

Figure 5.21 Incremental oil production in the Little Creek CO$_2$-flood pilot test (Stalkup 1983) ©SPE-AIME.

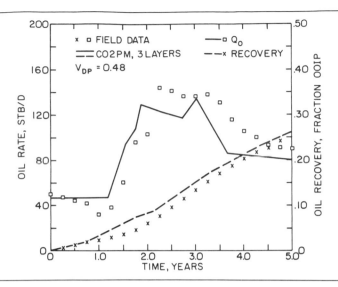

Figure 5.22 Comparison of CO_2 predictive model results with field data for the Slaughter Estate Unit Pilot (Paul, 1983).

Paul (1983) also compared the CO2PM with Shell Oil Co.'s WASSIM simulator for the Denver Unit, Wasson Field. Figure 5.23 shows that the CO2PM, run as a three-layer case, compares reasonably well with the Shell results. As in the Slaughter example, the CO2PM could not match the simulator unless layering was used.

5.5   **APPENDIX: MODIFIED KOVAL METHOD FOR INVERTED FIVE SPOTS.** The pertinent equations for the solution of oil recovery as a function of the pore volumes injected for continuous CO_2 injection as a secondary recovery fluid are given now. Koval's mixing coefficient to determine displacement efficiency is defined as

$$K = HFE, \tag{5.48}$$

$$K = HF[0.78 + 0.22 \, (\mu_o/\mu_{CO_2})^{1/4}]^4, \tag{5.49}$$

$$H = 1 \text{ (for homogeneous reservoirs)}, \tag{5.50}$$

$$\log_{10} H = [V_{DP}/(1 - V_{DP})^{0.2}] \text{ (for stratified reservoirs, see chapter 2)},$$

$$F = \text{gravity segregation correlation (see modified Koval method)}.$$

For breakthrough conditions,

$$(V_{pvd})_{BT} = 1/K, \tag{5.51}$$

$$(E_{vd})_{BT} = 1/K, \tag{5.52}$$

$$M = \mu_{oil}/\mu_{CO_2}, \tag{5.53}$$

$$(E_A)_{BT} = (1 + 0.4M)/(1 + M), \tag{5.54}$$

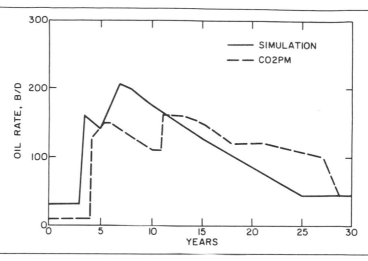

Figure 5.23 Comparison of CO₂ predictive model results and simulation results for the Denver Unit (Paul 1983).

$$(V_{pi})_{BT} = (E_A)_{BT} \cdot (V_{pvd})_{BT},$$ (5.55)

$$(V_{pa})_{BT} = (V_{pi})_{BT}/(E_{vd})_{BT}.$$ (5.56)

For linear displacements, E_A is set equal to one, and the reader should skip to equation (5.63) and solve directly for E_{vd} (remember for linear displacements that $V_{pi} = V_{pvd}$).

For five-spot pattern displacements, Claridge (1972) has developed a correlation for areal sweep. His results are shown in figure 5.24 with the following first five equations representing that work.

$$M_{BT} = [1 - (E_A)_{BT}]/[(E_A)_{BT} - 0.4]$$ (5.57)

$$M = 25[(M)_{BT}^{5/6} + 0.3 + 2.3(V_{pa} - 1)/(V_{pa} + 1)]$$ (5.58)

$$M = [M - M_{BT}/(\overline{M} - M_{BT})]^X$$ (5.59)

$$X = [0.85 - 0.55(E_A)_{BT} + 0.25V_{pa}]$$ (5.60)

$$E_A = [(E_A)_{BT} + 0.4M]/(1 + M)$$ (5.61)

$$V_{pvd} = \frac{(E_{vd})(V_{pa})}{(E_A)}$$ (5.62)

$$E_{vd} = \frac{2(KV_{pvd})^{1/2} - 1 - V_{pvd}}{K - 1}$$ (5.63)

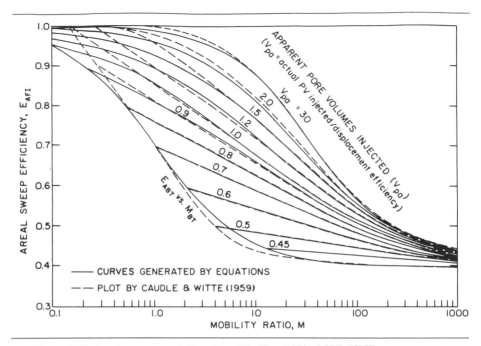

Figure 5.24 Areal sweep correlations from Claridge (1972) ©SPE-AIME.

$$V_{pi} = (V_{pa})(E_{vd}),\qquad\qquad(5.64)$$

where, by definition,

E_{vd} = volumetric displacement efficiency for a given well pattern (five-spot), vol/vol,

$(E_A)_{BT}$ = areal sweep efficiency at breakthrough per Caudle-Witte (1959) correlation,

E_A = areal sweep efficiency in the Caudle-Witte correlation corresponding to injection of V_{pa} HCPV of solvent,

V_{pa} = HCPV of solvent injection that corresponds to an apparent areal sweep in a five-spot pattern equal to that given by the Caudle-Witte correlation (E_a),

V_{pi} = HCPV of solvent injection in an ideal five spot of unit thickness that will produce an invaded area (containing fingers of solvent and intervening regions occupied by oil) whose outer envelope corresponds to E_A times the area of the five spot,

$(V_{pi})_{BT}$ = the value of V_{pi} at solvent breakthrough,

V_{pvd} = HCPV of solvent injection that, in an ideal linear system, will give a volumetric efficiency equal to E_{vd},

HCPV = an abbreviation representing hydrocarbon pore volumes—that is, volumes relative to the volume of hydrocarbons initially in place, which is taken to be equal to $Ah\phi(1 - S_{wi})$,

H = Koval's heterogeneity factor, vol/vol,

M = mobility ratio (for displacing fluid/displaced fluid), viscosity/viscosity,

K = Koval's modified mobility ratio, viscosity/viscosity,

E = Koval's effective viscosity ratio, viscosity/viscosity,

M_{BT} = mobility ratio of points on the solvent breakthrough curve in equations fitting Caudle-Witte's figure 4,

$\overline{M}$ = mobility ratio of the midpoint of the E_a versus M curve for a given value of V_{pa} in equations fitting Caudle-Witte's figure 4—that is, the mobility ratio at a value of $E_a - [(E_a)_{BT} + 0.4]/2$—for a given value of V_{pa},

F = gravity segregation correlation for override (see modified Koval method),

M = a function of M, $\overline{M}$, M_{BT}, $(E_A)_{BT}$, and V_{pa} used in fitting Caudle-Witte's figure 4.

Solution to the problem of solving for E_A and E_{vd} as a function of the actual pore volumes injected and ultimately the oil produced, N_p, for an inverted five-spot pattern becomes:

Using equations (5.48) through (5.56), estimate the areal sweep, E_A, the actual pore volumes, V_{pi}, injected at breakthrough, etc.

Make a list of apparent pore volumes injected above breakthrough, and determine E_A for each [eqs. (5.57) to (5.61)].

Continue by using equations (5.62) and (5.63) to determine the displacement efficiency, E_{vd}, and the idealized pore volumes injected, V_{pvd}.

Determine the actual pore volumes injected, V_{pi}, for each V_{pa}, E_A, V_{pvd}, and E_{vd} set, using equation (5.64).

Use equation (5.13) to determine oil production for each set of values.

REFERENCES
Blackwell, R. J., Rayne, J. R., and Terry, W. M.: "Factors Influencing the Efficiency of Miscible Displacement," *Trans., AIME* (1959) **219**, 1–8.

Carcoana, A. N.: "Enhanced Oil Recovery in Rumania," paper SPE/DOE 10699 presented at the SPE/DOE 3rd Joint Symposium on Enhanced Oil Recovery, Tulsa, April 4–7, 1982.

Caudle, B. H., and Witte, M. D.: "Production Potential Changes During Sweepout in a Five-Spot System," *Trans., AIME* (1959) **216**, 446–447.

Claridge, E. L.: "Prediction of Recovery in Unstable Miscible Flooding," *Soc. Pet. Eng. J.* (April 1972) 143–154.

Claridge, E. L.: "CO_2 Flooding Strategy in a Communicating Layered Reservoir," *J. Pet. Tech.* (Dec. 1982) 2746–2756.

Craig, F. F., Sanderlin, J. L., Moore, D. W., and Geffen, T. M.: "A Laboratory Study of Gravity Segregation in Frontal Drives," *J. Pet. Tech.* (Oct. 1957) 275–280.

Doscher, T. M., and El-Arabi, M.: "High Pressure Model Studies of Oil Recovery by Carbon Dioxide," paper SPE/DOE 9787 presented at the SPE/DOE 2nd Joint Symposium on Enhanced Oil Recovery, Tulsa, April 5–8, 1981.

Dumoré, J. M.: "Stability Considerations in Downward Miscible Displacements," *Soc. Pet. Eng. J.* (Dec. 1964) 356–362.

Dykstra, H., and Parsons, R. L.: *Secondary Recovery of Oil in the United States*, API, Washington (1950).

Geffen, T. M.: "Improved Oil Recovery Could Help Ease Energy Shortage," *World Oil* (Oct. 1973) 84–88.

Hawthorne, R. G.: "Two-Phase Flow in Two-Dimensional Systems—Effects of Rate, Viscosity and Density on Fluid Distribution in Porous Media," *Trans., AIME* (1960) **219**, 81.

Holm, L. W., and Josendal, V. A.: "Effect of Oil Composition on Miscible-Type Displacement by Carbon Dioxide," *Soc. Pet. Eng. J.* (Feb. 1982) 87–98.

Interstate Oil Compact Commission: *Determination of Residual Oil Saturation*, IOCC, Oklahoma City (1978).

Iyoho, A. W.: "Selecting Enhanced Oil Recovery Processes," *World Oil* (Nov. 1978) 61–64.

Khatib, A. K., Earlougher, R. C., and Kantar, K.: "CO_2 Injection as an Immiscible Application for Enhanced Oil Recovery in Heavy Oil Reservoirs," paper SPE 9928 presented at the SPE California Regional Meeting, Bakersfield, March 25–26, 1981.

Klins, M. A., and Farouq Ali, S. M.: "Heavy Oil Production by Carbon Dioxide Injection," *J. Can. Pet. Tech.* (Sept.–Oct. 1982) 64–72.

Koval, E. J.: "A Method for Predicting the Performance of Unstable Miscible Displacement in Heterogeneous Media," *Soc. Pet. Eng. J.* (June 1963) 145–154.

Lacey, J. W., Faris, J. E., and Brinkman, F. H.: "Effect of Bank Size on Oil Recovery in the High-Pressure Gas-Driven LPG-Bank Process," *J. Pet. Tech.* (Aug. 1961) 806–816.

Lewin and Associates, Inc.: "The Potential and Economics of Enhanced Oil Recovery," prepared under U.S. FEA Contract No. CO-03-50222-000, Washington, D.C. (April 1976).

Lewin and Associates, Inc.: "Economics of Enhanced Oil Recovery," prepared for U.S. DOE Contract No. DE-AC01-78ET12072, Washington, D.C. (May 1981).

McRee, B. C.: "CO_2: How It Works, Where It Works," *Pet. Eng.* (Nov. 1977) 52–63.

National Petroleum Council: *Enhanced Oil Recovery*, NPC, Washington, D.C. (Dec. 1976).

Office of Technological Assessment: *EOR Potential in the United States*, McGraw-Hill, New York (1978).

Patton, J. T., Coats, K. H., and Spence, K.: "Carbon Dioxide Well Stimulation: Part 1—A Parametric Study," *J. Pet. Tech.* (Aug. 1982) 1798–1805.

Paul, G. W. (principal investigator): "Development and Verification of Simplified Prediction Models for Enhanced Oil Recovery Application: CO_2 (Miscible Flood) Predictive Model," U.S. DOE Report DE-AC19-80BC10327, Intercomp-Denver (1983).

Pullman-Kellogg, Inc.: "Sources and Delivery of Carbon Dioxide for Enhanced Oil Recovery," U.S. DOE Report FE-2515-24, Washington, D.C. (1978).

Raimondi, P., and Torcaso, M. A.: "Distribution of the Oil Phase Obtained Upon Imbibition of Water," *Soc. Pet. Eng. J.* (March 1964) 49–55.

Reid, T. B., and Robinson, H. J.: "Lick Creek Meakin Sand Unit Immiscible CO_2/Waterflood Project," *J. Pet. Tech.* (Sept. 1981) 1723–1729.

Rowe, H. G., York, S. D., and Ader, J. C.: "Slaughter Estate Unit Tertiary Pilot Performance," *J. Pet. Tech.* (March 1983) 613–620.

Sankur, V., and Emanuel, A. S.: "A Laboratory Study of Heavy Oil Recovery with CO_2 Injection," paper SPE 11692 presented at the SPE California Regional Meeting, Ventura, March 23–25, 1983.

Shelton, J. L., and Schneider, F. N.: "The Effects of Water Injection on Miscible Flooding Methods Using Hydrocarbons and Carbon Dioxide," *Soc. Pet. Eng. J.* (June 1975) 217–226.

Stalkup, F. I., Jr.: "Displacement of Oil by Solvent at High Water Saturation," *Soc. Pet. Eng. J.* (Dec. 1970) 337–348.

Stalkup, F. I., Jr.: *Miscible Displacement,* Monograph Series, SPE, Dallas (1983) **8,** 63–96.

Taber, J. J., and Martin, F. C.: "Technical Screening Guides for the Enhanced Recovery of Oil," paper SPE 12069 presented at the SPE 58th Annual Technical Conference, San Francisco, Oct. 5–8, 1983.

Tiffen, D. L., and Yellig, W. F.: "Effects of Mobile Water on Multiple Contact Miscible Gas Displacements," paper SPE/DOE 10687 presented at the SPE/DOE 3rd Joint Symposium on Enhanced Oil Recovery, Tulsa, April 4–7, 1982.

Wang, G. C.: "Microscopic Investigation of CO_2 Flooding Process," *J. Pet. Tech.* (Aug. 1982) 1789–1797.

Warner, H. R.: "An Evaluation of Miscible CO_2 Flooding in Waterflooded Sandstone Reservoirs," *J. Pet. Tech.* (Oct. 1977) 1339–1347.

Welge, H. J.: "A Simplified Method for Computing Oil Recovery by Gas or Water Drive," *Trans., AIME* (1952) **195,** 91–98.

Yellig, W. F., and Metcalfe, R. S.: "Determination and Prediction of the CO_2 Minimum Miscibility Pressure," *J. Pet. Tech.* (Jan. 1980) 160–168.

6 PROJECT ECONOMICS AND DESIGN

The first two steps in the design sequence for implementing a CO_2 flood were discussed in chapter 5. They included a screening of the property to determine if field characteristics closely matched those of previously successful miscible or immiscible applications and a hand-held, rough estimate of field performance. This chapter deals with

continuing the prepilot evaluation (cost estimates and further data requirements),
a brief review of simulation methods for more accurate recovery estimates,
pilot testing and its objectives, and
final design.

6.1 PREPILOT EVALUATION

6.1.1 Project Economics. After preliminary screening and an attempt has been made to quantify oil recovery as a function of time, project economics must be identified. Normally, total process costs including taxes and capital costs for CO_2 floods are bounded between \$26 and \$39 per barrel of incremental oil recovered (Lewin and Associates 1981). When these financial and tax costs are neglected, CO_2 flooding costs (injectant plus investment and operating expenses) range between \$16 and \$27 per barrel of incremental oil (Taber and Martin 1983).

As shown in figure 6.1, CO_2 flooding costs compare most favorably with other EOR processes with perhaps one exception, steam flooding. However, steam is normally injected in thick, heavy oil zones and is not directly competitive with miscible displacements using CO_2. This is not true for CO_2-immiscible applications where pay thickness, depth, and CO_2 availability may prove to be the deciding factors between steam and CO_2.

Internally, the economics for CO_2 flooding are heavily front-end loaded, increasing the risk. Although specific costs will differ from project to project, an approximate

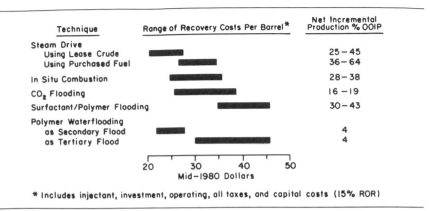

Figure 6.1 EOR cost range and recovery efficiency by technique (Lewin and Associates 1981).

breakdown (National Petroleum Council 1981) of the undiscounted cost elements *before taxes* includes

CO_2 purchase price	47%
CO_2 recycle operating expense	10%
Well operating cost	9%
Tangible investment including compressors, separation equipment, well conversions, and hydrogen sulfide removal	19%
Tangible and intangible drilling costs*	2%
Fuel cost	13%
	100%

In their *after tax* analysis, Bloomquist et al. (1981) found that the three tax components—windfall, federal, and severance—consumed an average of over 40% of the net future revenue. Additional work by Lewin and Associates (1981) also showed taxes to be over 30% of the cost per barrel of oil produced. These studies lead one to conclude that CO_2 price, oil price, and taxation rate will be the driving economic forces in a successful CO_2 flood.

Conceptually, the economics of a CO_2 injection project are computed with a revenue, cost, operating income, tax, and profit balance sheet on an annual basis as follows:

Project revenue   (Year)
Net oil sold (less royalty)
Net gas sold (less royalty)   *Total revenue (R)*

Project costs
Fixed operating cost
Variable operating cost

*Note that this assumes few or no new wells are required.

CO_2 purchase cost
CO_2-treating and reinjection cost
Produced water-treating and injection cost   — *Total operating cost (O)*

Capital costs
Well workovers
New wells (as required)
CO_2 injection distribution system
CO_2 delivery pipeline
Water injection plant (as required)
CO_2 recycling plant   — *Capital cost (C)*

Taxes
Windfall profit tax
State severance tax
Other state and local taxes
Federal and state income taxes   — *Tax cost (T)*
   Annual after tax cash flow *(P)*

That is, $P = R - O - C - T$, with the sum of the after tax cash flows, P, discounted to present value defined as the after tax DCF (discounted tax flow), sometimes called *present value profit*. The value of DCF for a project and its confidence interval (upside potential and downside risk) are usually the primary criteria upon which a project is judged. Economic criteria such as P/I ratio, \$/bbl profit, \$/bbl cost, and ROR provide additional sensitivity and decision risk information. However, the primary indicator of project incremental economics is usually DCF.

Table 6.1 shows the cost categories that must be estimated individually for a typical CO_2 project. The development of accurate cost data for all entries in this table is a major effort, requiring preliminary designs for CO_2 injection, surface production, water injection, and CO_2 recycling systems. For screening purposes, however, estimates applying uncertainty can be made. Major tangible investment costs to be estimated are

     new wells that are required,
     pipeline lateral to CO_2 trunkline or source,
     water injection plant or improvements to existing plant, and
     CO_2 recycling plant.

In a report by Lewin and Associates (1981), cost data for new wells and pipelines are reported for U.S. petroleum-producing regions. Table 6.2 gives estimates of new well-drilling and completion cost as a function of depth along with other costs. These other costs include production and injection expenses, workover costs, and annual operating expenses.

Transporting CO_2 to the field normally can be accomplished by one of three methods: tank truck, railcar, or pipeline. In most cases, tank truck and railcar transport systems should prove more expensive than pipelining. Since the flow of CO_2 could be curtailed by a labor shortage or a shortage of tank trucks or railcars, pipelining appears to be the most economical and reliable method of CO_2 transportation.

Table 6.1 Cost Categories for CO_2 Miscible Flood Projects

Cost Category

Capital costs
Feasibility study, lab work, project engineering and design, purchasing and interest during
   construction
CO_2 delivery/resale pipeline
Injection well workovers
Producing well workovers and installation of high capacity lift systems
New development wells
Injection distribution system
Surface production facilities
Water injection plant including filtration and treating
CO_2 recompression plant including dehydration

Operating costs
Fixed annual field operating cost for surface production facilities and direct overhead—i.e., basic
   cost to sustain operation
CO_2 delivery/resale pipeline
Fixed well operating cost
Unit lifting, treating, and handling cost for produced fluids
      Oil
      Water
Water treating and injection plant
CO_2 recompression/dehydration plant

Source: Intercomp Consortium Report, "Feasibility of CO_2 Enhanced Oil Recovery in the Rocky
Mountain Region," Intercomp, Denver (1980).

Table 6.2 Cost Equations, By Component and Area (Lewin and Associates 1981)

Area	Cost Equation (mid-1980 dollars)
Drilling and completion	
2	$48,451\ e^{(.00032D^a)}$
2A	$128,390\ e^{(.00032D)}$
3	$55,335\ e^{(.00027D)}$
4	$51,688\ e^{(.00028D)}$
5	$30,392\ e^{(.00034D)}$
6	$30,430\ e^{(.00035D)}$
6A	$688,514\ e^{(.00011D)}$
7	$29,360\ e^{(.00035D)}$
8	$45,167\ e^{(.00038D)}$
9	$23,742\ e^{(.00039D)}$
10	$16,257\ e^{(.00051D)}$
11A	$149,329\ e^{(.00030D)}$
Production equipment	
1, 2, 2A	$33,392\ e^{(.00011D)}$
3, 4	$21,509\ e^{(.00015D)}$
5 to 11A	$24,908\ e^{(.00014D)}$

Table 6.2 (continued)

Remaining lease equipment	
1, 2, 2A	$34{,}797\ e^{(.00003D)}$
3, 4	$24{,}027\ e^{(.00003D)}$
5 to 11A	$20{,}183\ e^{(.00004D)}$
Injection equipment	
1 to 11A	$22{,}892\ e^{(.00009D)}$
Annual operation	
Primary recovery	
1, 2, 2A	$7{,}567\ e^{(.00009D)}$
3, 4	$7{,}290\ e^{(.00006D)}$
5 to 11A	$7{,}292\ e^{(.00006D)}$
Enhanced recovery	
1, 2, 2A	$14{,}139\ e^{(.00013D)}$
3, 4	$13{,}283\ e^{(.00011D)}$
5 to 11A	$13{,}298\ e^{(.00011D)}$
Workover	
1 to 11A	0.48 drilling and completion + 0.50 production equipment
Drilling and production area location	
1	Alaska
2	California, Oregon, Washington
2A	Pacific Coast (offshore), Idaho, Nevada, Utah
3	Colorado, Arizona, New Mexico (west)
4	Wyoming, Montana, North Dakota, South Dakota
5	Texas (west), New Mexico (east)
6	Texas (east), Arkansas, Louisiana, Mississippi, Alabama
6A	Gulf Coast (offshore)
7	Oklahoma, Kansas, Nebraska, Missouri, Iowa, Minnesota
8	Michigan, Wisconsin
9	Illinois, Indiana, Kentucky, Tennessee
10	Ohio, Pennsylvania, West Virginia, New York, Virginia, North Carolina
11A	South Carolina, Georgia, Florida

[a]D = depth (ft).
Source: Lewin and Associates, Inc., "Economics of Enhanced Oil Recovery," report DOE/ET/78–C–01–2628 (March 1981).

The pipeline system that will carry the CO_2 from a natural CO_2 well site or production plant can be operated under one of three different pressure conditions:

1. A subcritical pipeline system (line pressure is less than the critical pressure of CO_2, 1,071 psia),
2. A supercritical pipeline system (operating pressure is greater than 1,071 psia),
3. A liquid pipeline system.

Work by Pullman-Kellogg (1978) concluded that the supercritical CO_2 pipeline system was the most economical system available in transporting large quantities of CO_2 for enhanced oil recovery.

To estimate the pipeline costs per mile at any capacity, the following equation may be used (Lewin and Associates 1981):

$$\text{Unit pipeline cost (\$/mi)} = 100,000 + 2,008[(\text{MMSCF/D})^{0.834}] \qquad (6.1)$$

Note that this equation was derived for 8- to 18-in pipeline diameters with flow rates from 50 to 500 MMSCF/D.

The additional effect of terrain on per mile costs is summarized by Pullman-Kellogg (1978):

$$\text{Rolling hill: (\$/mi)} = 13,350 + 2.8(\text{MMSCF/D}),$$

$$\text{Rugged terrain: (\$/mi)} = 118,750 + 437.5(\text{MMSCF/D}),$$

$$\text{River crossings: (\$/mi)} - 131,250 + 512.5(\text{MMSCF/D}).$$

Again, these cost estimates should only be used for pipeline design rates of 50 to 500 MMSCF/D.

In order to estimate the size pipeline needed, table 6.3 is presented. Once the size of the field to be flooded has been determined, pipeline capacity and CO_2 transportation costs can be estimated. This transportation price includes a recoupment of the capital outlay for the pipeline at a pretax rate of 20% and does not include the purchase cost of the CO_2. It then becomes a simple matter to determine which size pipeline (diameter) will perform for the selected flow rate and length given.

Both water injection and CO_2 recycling plants require considerable analysis to deter-

Table 6.3 Estimating Pipeline Capacity and CO_2 Delivery Cost

Estimated Pipeline Capacity (MMSCF/day)	Minimum Required Field Size		Estimated CO_2 Delivery Cost (mid-1980 \$/MSCF/100 mile of pipeline)
	Incremental Oil Recovery by CO_2 (in million barrels)	Residual Oil in Place (in million barrels)	
300	146	730	0.09[a]
200	98	490	0.10
100	48	240	0.14
50	24	120	0.20
25	12	60	0.31
10	6	30	0.56
5	4	17	1.12

Source: Lewin and Associates, Inc., "Economics of Enhanced Oil Recovery," report DOE/ET/78–C–01–2628 (March 1981).
[a]This price only includes a recoupment of pipeline investment costs (20% pretax DCFROR) and transportation charges, not the actual cost of purchasing CO_2.

mine estimated investment cost. However, for a typical design, the water injection plant cost may range from $750,000 at 10,000 bbl water/d to $3,000,000 at 50,000 bbl water/d. A typical CO_2 recycling plant design price can range from $1,500,000 at 10 MMSCF/D to $5,000,000 at 50 MMSCF/D. Depending on the need for recompression or gas separation in the project, the investment cost might be increased considerably (Intercomp Report 1980).

Costs of the CO_2 vary widely. For natural CO_2 cost may run from $0.40 to $2 per MSCF at the wellhead, while manufactured CO_2 costs could be 70% to 100% higher.

Recycling CO_2 during the life of the project will usually be advantageous relative to purchasing the total amount of CO_2 to be injected. The recycling operation will include dehydration and recompression of produced gas and may or may not include separation of CO_2 from the other gases in the production stream. Cost estimates for recycling include $0.14/MSCF for hydrocarbon separation, $0.14/MSCF for H_2S separation, and $0.26/MSCF for repressuring (Lewin and Associates 1981).

6.1.2   **Further Data Requirements.** In the design of a CO_2 flood, the candidate reservoir must first pass preliminary screening criteria with a review of the potential economics. Extensive data gathering need now be done in the field and laboratory before more sophisticated numerical reservoir simulators can be employed. Input data common to all simulators (black oil, compositional, miscible, and hybrid) and their various sources are listed in table 6.4.

As discussed in chapter 3, the need for various laboratory-derived fluid property data will depend on the type of process (immiscible, multiple-contact miscible, or assumed

Table 6.4 Reservoir Simulator Input Data versus Reliability

Parameter	Source
Production-injection data	Field tests and measurements
Producing and injecting bottomhole pressures	Field tests and measurements
Fluid property data	Laboratory PVT analysis
Net pay thickness	Pay studies: log, core, pressure transient, and performance data
Connate water saturation	Log and core analysis
Gas, oil, and water relative permeability data	Laboratory core analysis
Reservoir pressure	Material balance calculations, pressure build-up data
Initial gas saturation	Material balance calculations, waterflood, k_g/k_o data
Effective wellbore radii and condition ratios	Pressure transient data
Interwell k_h	Pressure transient data, core analysis
Stratification	Core analysis, performance

Source: Drennon et al., "A Method for Appraising the Feasibility of Field-Scale CO_2 Miscible Flooding," paper SPE 9323, Dallas (1980).
Note: List is from highest to lowest reliability.

first-contact miscible) to be employed and the type of model used to predict that process's performance. Typical information that might be required for sophisticated numerical simulation includes

1. minimum miscibility pressure,
2. crude oil swelling and viscosity reduction,
3. single and multiple contact PVT data, and
4. asphaltene precipitation.

Immiscible displacements may only require information about crude oil swelling, viscosity reduction, and asphaltenes; first-contact miscible tests might also include minimum miscibility experiments. Multiple-contact processes will require the full suite of laboratory experiments.

The experiments run to provide this information, most often, fall into three general categories:

1. High pressure volumetric (PVT) and vapor/liquid equilibrium (VLE) experiments,
2. Slim tube displacements, and
3. Core displacements.

A thorough examination of these typical tests and their importance was made in chapter 3.

In review, Orr et al. (1982) presented the following conclusions concerning laboratory experiments and their efforts to model CO_2-oil interaction in the reservoir:

The suite of laboratory experiments used to evaluate a particular field for CO_2 flooding should be chosen according to the characteristics of the field and the type of numerical simulation to be used for field-scale recovery predictions.

Because they eliminate effects of viscous fingering, gravity segregation, and rock heterogeneity, slim tube displacement experiments offer the simplest method for evaluating the effectiveness of the phase behavior portion of the displacement process.

Core floods do not always eliminate such effects and, hence, must be interpreted with care.

Equilibrium phase behavior experiments provide data useful for tuning phase behavior calculations in compositional reservoir simulators but are time consuming and expensive.

6.2   **NUMERICAL SIMULATION.** A matched set of laboratory/field data with the proper simulation technique must be made for the final design of any large-scale CO_2 flood. The simulator is normally tuned to match past field history, be it waterflooding or natural depletion, by altering the reservoir properties with which the operator has the lowest confidence. These usually include initial permeabilities, porosities, and gas saturation.

Checkpoints in evaluating the quality of a history match include water cut, peak oil rate, oil response time, and water breakthrough time for a waterflood history match.

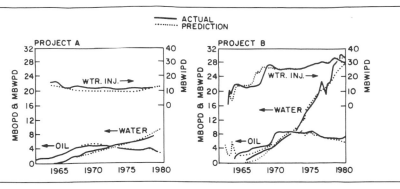

Figure 6.2 Typical waterflood history matches (Drennon et al. 1980).

Work by Drennon et al. (1980) discusses such a history match procedure. Figure 6.2 illustrates their history matching of waterflood performance prior to using the simulation model as a CO_2 prediction tool.

The models described in detail in chapter 4 and specifically applicable to CO_2 flooding fall into three general groups:

1. Black oil immiscible simulators,
2. Compositional simulators, and
3. Black oil mixing parameter miscible simulators.

Also, as discussed in chapter 4, there are combination models that synthesize miscible and compositional techniques.

The basic difference between the three different types of simulation models—compositional, black oil immiscible, and miscible—lies in their treatment of fluid and phase properties. It is important, therefore, to choose not only a simulator whose design characteristics match the transport mechanisms occurring in the reservoir but also, as discussed earlier, a set of laboratory experiments that fits model requirements and that will reduce the amount of data that must be estimated to conduct numerical simulation.

The assumptions involved in compositional models are multiphase flow, two or more components, and negligible dispersion. The assumptions in black oil immiscible models are multiphase flow (gas, oil, and water); up to three components called gas, oil, and water; and negligible dispersion. Finally, a miscible model assumption involves a single phase, usually two components, and significant dispersion.

6.2.1 **Black Oil Immiscible Model.** For CO_2 injection, black oil (beta-type) simulators may be adequate in predicting performance of low pressure displacements. This process of immiscible CO_2 injection occurs in the pressure-temperature area approximated by Region I on figure 6.3.

Black oil models are capable of simulating systems where water, oil, and CO_2 are present in all proportions. However, if a fourth fluid, natural gas, is present in signifi-

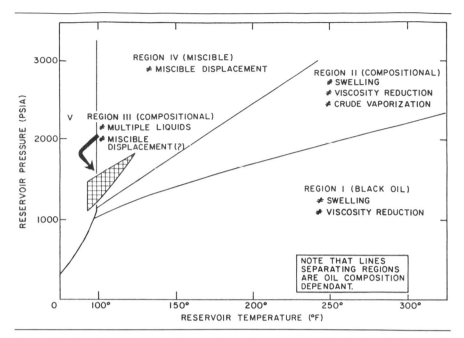

Figure 6.3 The effect of reservoir temperature and pressure on CO_2 injection displacement mechanisms.

cant quantities, this method may fail due to its inability to account for mixing of gases. Usually, phase transfers are accounted for between the gas and water and the gas and oil phases by solution gas-oil-water ratio functions. There is no vaporization of liquid components.

Fluid properties in the beta-type simulator are all obtained from the oil formation volume factor, water formation volume factor, gas formation volume factor, phase viscosities, solution gas-oil ratio, and solution gas-water ratio, all specified functions of individual phase pressure only. These data are easily obtained by routine reservoir-fluid volumetric and fluid property measurements in addition to standard core characterization (permeability, relative permeability, and capillary pressure).

6.2.2   **Compositional Model.** When there is considerable mass transfer between the flowing phases, use of a multicomponent, compositional model is dictated. This type of phase transfer is typical of the vaporization or condensation processes taking place in Region II, Region III, and the lower section of Region IV on figure 6.3. Multiple-contact miscibility processes like these cannot be formulated accurately on black oil simulators.

However, the compositional model is most complex since its focus is on individual components rather than phases. An EOS must be relied upon to take block compositions and to predict effectively the PVT behavior of the mixture. Thus, considerable input data are needed, in addition to the usual reservoir description data, to define the composition of the reservoir fluids and to specify the parameters used in the computations for phase compositions and fluid properties. This time-consuming process of determining the EOS interaction parameters usually involves single- and multiple-contact phase equilibrium experiments and the defining of oil pseudocomponents for trial and error matching of the experimental data. Also, there is a significant increase in the number of variables, which raises computational time requirements.

Unlike the black oil model, where phase properties (μ, β, ρ, etc.) are functions of phase pressures, in the compositional model phase properties are calculated as strong functions of phase composition as well as the phase pressure and reservoir temperature. To date, the best predictive methods of phase composition are based on the Redlich-Kwong (1949) EOS and include Wilson (1969), Zudkevitch and Jaffe (1970), Soave (1972), Peng and Robinson (1976), and Yarborough (1978) modifications (see section 3.9).

Lastly, compositional as well as black oil immiscible models neglect dispersive mass transfer as a rule.

6.2.3   **Black Oil Mixing Parameter Miscible Model.** Because of the complexity of the solution of a set of generalized compositional equations, there was strong motivation to develop miscible simulation models using the conventional black oil model premise. Early work by Koval (1963) pioneered the way in predicting miscible displacement performance long before the advent of numerical simulation. His method, analogous to the Buckley-Leverett method, predicted oil recovery and solvent cut as a function of the pore volumes of solvent injected.

In 1972, Todd and Longstaff (1972) extended Koval's as well as a number of other authors' work in developing a black oil simulator for predicting miscible flood performance. They introduced the mixing parameter, ω, to determine the fluid properties, μ and ρ, of the dispersed zone between solvent and oil. This feature allowed the use of coarser grids in reservoir simulation without masking the effects of dispersion (viscous mixing) at the flood front.

A miscible-type model (Koval or Todd-Longstaff) involves a simplification of phase properties. The fluid is always assumed to be miscible on contact and was described earlier as first-contact or very rapid multiple-contact miscibility (high in Region IV of fig. 6.3). This may be true when actual displacement pressures in the reservoir will be much higher than the predicted minimum miscibility pressure.

Also, the density and viscosity of the mixture depend primarily on solvent concentration, and the formulation assumes no volume change on mixing. Explicit treatment of dispersion makes miscible models useful in situations where recovery is controlled by sweep efficiency, and the interaction of viscous fingering and dispersion plays an important role.

If you have a process that is completely miscible on first contact (unlikely for CO_2) or in which the distance required to obtain miscibility is short with respect to well spacing, the major problem in the application of this scheme lies in the selection of the appropriate value of the dispersion mixing parameter; the Todd-Longstaff ω or Koval's K.

In review of simulation models, Orr et al. (1982) have noted that none of the simulators currently available models all the factors that are known to influence the efficiency

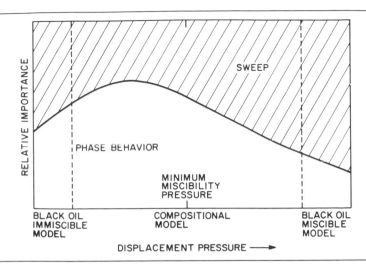

Figure 6.4 Relative importance of sweep efficiency and phase behavior as a function of displacement pressure.

of the process; therefore, all have limitations that should be considered when simulation results are evaluated (see fig. 6.4). Miscible simulators ignore phase behavior but attempt to model the effects of viscous fingering. Compositional simulators model phase behavior but do not account for viscous fingering and are affected by numerical dispersion, which alters the calculated composition path and, hence, the calculated process efficiency. Black oil immiscible models are applicable to a small number of CO_2 displacements and cannot account for gas in place. Nevertheless, there is really no alternative to the use of numerical simulation for CO_2 flood design. It simply is not possible to perform experiments in which all the relevant variables have been scaled properly.

Once a simulation type is chosen, the data developed, and a field match prepared, the mathematical model becomes a powerful (and irreplaceable) tool for extending and interpreting laboratory test results for field applications. The model will permit parametric screening of potential operating conditions (injection rate, slug size, gas/water injection ratio, pattern size, etc.) leading to what one hopes is an optimal design for the proposed flood.

6.3   **PILOT TESTING.** Before implementing a CO_2 flood or any EOR project on a fieldwide basis, a miniflood should be performed. The usual objectives in conducting a pilot application in the field are threefold:

1. To review the process design as determined from simulation results,
2. To gain further information about formation characteristics,
3. To examine potential operating problems that may arise.

6.3.1   **Design Analysis.** Prior to field testing, parametric studies with numerical simulators are used extensively in determining injection procedure, predicting timing of recovery, and estimating fluid-handling volumes. The pilot test can then be used to review the following:

Injectivity: Can projected injectivity be attained?

Slug size: Does the optimum sized slug mobilize oil in significant quantities? If not, can increased size or injection pressure improve results?

Mobility control: In horizontal floods where water may be injected simultaneously or in a WAG fashion, does the water/gas ratio appear to yield adequate mobility control? Does the injected water shield the CO_2 from contacting the oil, and is water/gas gravity segregation a problem? For vertical displacements where the CO_2 gas may be chased by another less costly gas like nitrogen, is the vertical sweep (a function of displacement velocity/critical velocity) adequate?

Preflush: A preflush of water may be used to raise reservoir pressure up to projected miscibility pressure prior to CO_2 injection. This is usually water injected ahead of the CO_2 in a horizontal displacement or at the oil/water contact in a vertical displacement. Is it feasible?

Timing of recovery: A pilot test does not yield incremental oil recovery results that can be extrapolated to the field. A single-pattern pilot test is neither confined (fluid may cross pattern boundaries and not be recovered) nor can it be assumed to represent field properties in its entirety. However, pilot tests give qualitative results and reassurance that oil was mobilized.

Recycling gas: How much CO_2 must be purchased until produced volumes are adequate for reinjection?

Separation systems: What type of production rates will be expected for sizing liquid-liquid, liquid-gas, and gas-gas separation, dehydration, and compression facilities?

6.3.2   **Formation Analysis.** Prior to pilot testing, a reservoir description is usually pieced together from available log and core data. Without performance tests, little confidence can be placed on some of this information. To provide a more accurate description of the reservoir for later simulation efforts, many tests including relogging, recoring, pressure transient tests, and tracer studies should be included in the pilot test package:

Coring: Initial oil saturation;

Drill stem testing: Reservoir pressure;

Pressure buildup: Formation flow capacity and skin damage;

Log-inject-log: Initial oil saturation;

Injection profiling: Temperature and radioactive logging for vertical conformance;

Pressure fall-off: Reservoir pressure, average Kh, and effective wellbore radii;

Pulse tests and interwell tracers: Directional flow, barriers, or anomalies. Tracers include tritium, IPA, ammonium thylocyanate, ammonium nitrate, or sulfur hexafluoride.

6.3.3   Operating Problems Analysis. Selection of proper materials and equipment will help to side step many of the operating problems associated with CO_2 injection. This section deals briefly with specific problems that may arise during a CO_2 flood and some of the precautions that can be taken, including the following:

Corrosion control: CO_2 and H_2O react to form carbonic acid. Proper dehydration and separate CO_2 and H_2O injection lines will allow common carbon steel pipe to transport CO_2 throughout the field. However, CO_2-H_2O mixing in WAG-type injection wells and in all production wells can lead to serious corrosion problems. Besides dehydration, corrosion control can include plastic-coated or cement-lined tubing (problems have been reported with collapsed plastic liners) and valves, stainless steel wellheads and valves, inhibitors, and oxygen scavengers. Monitoring equipment includes corrosomitor probes, hydrogen probes, corrosion coupons, corrosion test nipples, and ultrasonic metal thickness detectors.

Asphaltenes: Contact of high asphaltic crudes by weak acid may precipitate out a tarlike substance known as asphaltenes. Simple laboratory tests mixing and filtering resident crude oil with CO_2 and/or carbonated water should be run prior to field injection.

Dehydration of injection gas: Again, dehydration of the injected CO_2 reduces the acidity of the H_2O/gas mixture and also the potential for hydrate formation. Standard solutions to the problem include flowing the gas over molecular sieve material, silica gel, or glycol (EG, DEG, or TEG).

Separation of CO_2-H_2S-hydrocarbon gas mixtures: Commercially available processes to purify CO_2 prior to reinjection are amine scrubbing, cryogenic separation, molecular sieve material, or physical absorbant material such as Selexol, Rectisol and Sulfinol (Kuuskraa et al. 1981). It is likely that the operator, for safety reasons, will separate any H_2S produced with CO_2 prior to reinjection even though H_2S lowers the minimum miscibility pressure. Separation of the methane and ethane fractions will depend on the volumes produced. Significant quantities of CH_4 comingled with CO_2 production may be sold rather than reinjected. Also, as discussed in chapter 5, methane contamination increases miscibility pressure.

Additional operating problems: Other problems have been recorded that might dictate the use of CO_2 service-rated production packers and pumps. Deposits including sulfur and iron oxide as well as calcium carbonate scale and paraffin plugging may occur. Also, choke freezing has been reported as produced CO_2 is throttled from the wellhead. Larger chokes and a higher back pressure may be required.

Every pilot and full field design will not encounter all the foregoing problems. However, for a test to be successful, careful planning at every crossroad must be taken (see Adams and Rowe 1980; Graham et al. 1980; Stalkup 1983). Otherwise, valuable information will be lost and poor design will result with an even worse performance following.

6.3.4   Pilot Implementation. The size of the pilot project is usually determined by the design life of the pilot. Pattern size should provide substantial oil response within two to three years of commencing CO_2 injection. Too large a project could drawn out the pilot's life

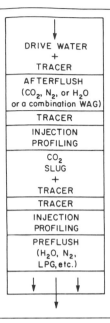

Figure 6.5 Schematic showing a recommended procedure for injecting CO₂ pilot test fluids (Mungan 1981).

to an economically unacceptable length. Too small a pattern size may mask true performance if isolated heterogeneities are contained in the flooded zone.

An injection-ordering scheme for pilot testing of the CO_2 process has been suggested by Mungan (1981) in figure 6.5. The preflush, usually water, raises the pressure in depleted reservoirs to the minimum miscibility pressure. Injection profiling may then be carried out to determine the volume and distribution of the fluid intake by layer. Tracers are injected next to examine interwell communication and sweep efficiency. Injected next is the CO_2 bank with a tracer. This tracer can be used to compare its movement versus the CO_2's movement, yielding information on directional permeability and volumetric sweep. Injectivity profiling and tracers following the CO_2 bank allow comparison with performance prior to CO_2 injection. The final drive fluid should also utilize tracers to determine final breakthrough.

Mungan (1981) has pointed out that not every field application will involve all the foregoing steps. Individual operators may add or delete steps, depending on the type of information—for example, injection profiling, sweep efficiency, breakthrough timing—that is required. The important thing to remember is that the information gained by pilot testing, an increased confidence that incremental oil will be recovered while overcoming potential operating questions, usually justifies its added expense.

However, contrary to all that has been discussed to this point, economic factors may not justify pilot testing in some cases. If laboratory and simulation studies are overwhelmingly favorably, if there is significant experience in similar field types with the same process, or if delay of fieldwide application cannot be economically justified, pilot testing may not be warranted.

In one example, Christian et al. (1981) planning a nitrogen flood in their Jay-Little Escambia field, stated that while a pilot test could yield useful information regarding recovery, it had three strong disadvantages: (1) implementation of the fieldwide project would be delayed at least five years, (2) cost of the pilot would be $7 million, and (3) they expected a pilot to be difficult to interpret.

As an intermediate alternative between whether or not to pilot, a CO_2 minitest might be employed. The design, implementation, and evaluation of one such minitest is discussed in detail by Desch et al. (1982). Their test, conducted in the Mission Canyon Formation at Little Knife, North Dakota, was a *nonproducing*, 5-acre field trial located in the center of the field. The test pattern consisted of a single injection well offset by three observation wells in an inverted four-spot configuration and involved time-lapse logging and sampling to monitor saturation changes as alternate slugs of water and CO_2 passed the observation wells. A fourth observation well was drilled to pressure core the project interval within the 5-acre pattern, and the resulting saturations were used for comparison with log saturations. Their minitest was then designed to give the following information: reduction in original oil saturation due to water injection, reduction in waterflood residual oil saturation due to alternate CO_2/water injection, extent of gravity segregation, effect of stratification and cross flow, and influence of reservoir heterogeneity.

The WAG injection sequence was completed within 10 months with tracers in the preflush and WAG cycles. Results were then history matched (bottomhole pressure, water, and breakthrough times for the three observation wells) with a compositional simulator, and the resulting information was extended to examine the potential of a future fieldwide project.

6.4   FIELD TESTS

6.4.1   Miscible Applications. Over 40 miscible design CO_2 injection projects have been initiated worldwide since 1958 when carbonated water was injected into the K&S Project of northeastern Oklahoma. Many of these projects are listed in table 6.5.

Mead Strawn was the first pure CO_2 slug project (Holm and O'Brien 1971). Started in 1964, this pilot project was conducted in the Mead Field near Abilene, Texas. The process consisted of injecting a preflush of water to raise reservoir pressure to 850 psig, followed by a small slug of CO_2 (4% PV), chased by a slug of carbonated water (12% PV) and then brine. The Upper Strawn Sand averages 9 ft of net thickness and 9.4% porosity at a depth of 4,475 ft. Original oil saturation was 60% with a permeability to air of 9 md. Results showed that over 50% more oil was produced by the CO_2-carbonated waterflood than by conventional waterflooding. Although a technical success, low permeability caused an extended flood life and adversely affected the economics of the process.

Nine years later, in 1972, the largest CO_2 and enhanced recovery project in the world was initiated (Kane 1979). The Kelly-Snyder Field in Scurry County, Texas, containing 2.73 billion bbl of oil originally in place, is one of the major oil reservoirs in the United

States. In 1953, the SACROC (Scurry Area Canyon Reef Operating Committee) unit reverted to a waterflood and performance was encouraging, but technical representatives were investigating potential methods for improving ultimate oil recovery above a projected 20%. In 1972 a WAG (CO_2) process was instituted.

There are now three project areas totaling 33,000 acres overlain by 160-acre inverted nine spots. A small amount of water was preinjected to raise pressure to the miscibility level, and alternating slugs of CO_2 and water followed. Continuous water now trails the 12 to 15% HCPV CO_2 slugs. Projected ultimate oil recovery is expected to be approximately 200 million bbl over and above that from the original water injection program. These results are remarkable considering that the 700 ft of heterogeneous productive zone averages 210 ft of net thickness with low permeability.

The Crossett Field (unitized as the North Cross Unit in 1964), in west Texas, is a chert-limestone pay characterized by a high porosity, averaging 22%, and low uniform permeability, 5 md (Pontious and Tham 1978). Primary production by solution gas drive was estimated to recover 12.9% OOIP. Secondary recovery operations using waterflooding were not feasible because of extremely low permeabilities. As an alternative, CO_2 was purchased from the nearby SACROC CO_2 pipeline and injected. Seventy-three BSCF of gas (40% HCPV) are planned to be injected in the six injection wells covering 1,700 acres. No preflush or chase water was employed, however; produced gas is now reinjected in the gas cap.

After six years of injection, cumulative production totaled 9.6 MMSTB of oil with an estimated incremental oil recovery due to CO_2 of 926 MSTB. This is slightly over 1% of the estimated 70 MMSTB of oil originally in place. Pontious and Tham (1978) report that while the production increases at the North Cross Unit have not been as high as original predictions, response has been definite and encouraging. Evidence to date indicates that the displacement mechanism is very complex, probably more complex than predicted by numerical simulation. The results thus far are encouraging, and the project is continuing.

A tertiary recovery process pilot in the Slaughter Estate Unit was initiated in August 1976 (Rowe et al. 1982). An alternate solvent gas (72% CO_2 and 28% H_2S) and water injection scheme showed production response in October 1977. Nitrogen gas (29.6% HCPV) will chase the 26% HCPV of solvent gas already injected. These gases will be trailed by H_2O.

The 12.9-acre Slaughter Estate Unit pilot is located in Hockley County, west Texas, and is produced from the moderately oil-wet San Andres formation. Pay properties include a depth of 4,950 ft, 75.2 ft of net pay, with a porosity of 10% and a water saturation of 8.1%. Cumulative incremental tertiary oil production after five years of injection was estimated to be 95,680 STB, which represents 14.9% of the original oil in place. Simulation efforts have shown that the final incremental tertiary recovery will be 20% to 25% OOIP. This means that cumulative recovery at the end of CO_2 injection should surpass 70% OOIP.

Last, the Weeks Island S Sand salt dome reservoir CO_2 pilot flood tested a gravity stable displacement of a 29.3% residual oil saturation downdip (32°). The reservoir is at 12,700 ft, making it one of the deepest CO_2 pilots to be conducted. Reservoir temperature is 225°F with a miscibility pressure of 5,100 psia (Perry 1982).

An 860 MMSCF slug of CO_2 cut with 5% natural gas was injected at the gas-oil contact and maintained by reinjection of the produced gas and replacing the downdip

Table 6.5 Summary of CO_2 Field Tests (Miscible Design)

Field	Formation	Depth, ft	BHT, degrees F	Oil Gravity, degrees API	Oil Viscosity, cp	Permeability, md	Porosity, %	Sand Thickness, ft Gross	Net	Miscibility Pressure, psi	Field Spacing, Acres/Well	Data Test Initiated	Field Productive Acres
Alabama													
Citronelle Unit	Rodessa-SS	11,300	—	44	—	0.1 to 400	12 to 20	—	—	—	90	1980	13,640
California													
N. Cole Levee, Kern County	Stevens-SS	9,206	235	32	0.28	0.5 to 1,200	15.9	—	—	—	9	1981	70
Colorado													
McCallum Unit, Jackson	Pierre "B"-SS	1,250	70	20	10	98	20	195	—	1,600	32	1973	601
Louisiana													
Bay St. Elaine	8,000-SS	7,400	164	36	0.7	1,480	33	100	35	3,300	—	1981	10
Paradis, St. Charles	No. 8-SS	9,600	192	39	0.39	720	28	—	—	—	29	1981	260
Paradis, St. Charles	Lower 9,000'-SS	10,400	204	33	0.5	663	25	—	—	—	20	1981	234
Quarantine Bay	4 Sand Res C-SS	8,075	190	32	0.99	100 to 1,000	29	—	—	3,200	8	1981	47
Weeks Island, Theris	S Sand	12,700	225	32	0.3	1,800	26	210	150	5,100	5	1978	679
Mississippi													
Little Creek, Lincoln Co.	Tusc-SS	10,400	248	30	0.4	75	23	360	29	5,000	40	1974	6,310
Tinsley	Perry-SS	4,900	60	39	1.5	65	26.4	—	—	—	62	1981	1,120
New Mexico													
Maljamar	Grayburg-SS/San Andres-Dol	4,000	91	35	1.1	30	11.7	—	—	—	2	1980	5
North Dakota													
Little Knife, Billings	Mission Canyon-Dol	9,800	240	43	0.2	22	18	230	120	3,000+	5	1979	5
Oklahoma													
Garber, Garber Co.	Crews-SS	1,950	—	45	1.7	35	17	—	—	1,100	6	1981	85
Purdy Springer, Garvin Co.	Springer A-SS	9,400	—	38	0.9	44	13	—	40	2,100	47	1982	4,960
East Velma, Garfield Co.	Sims C-2 SS	7,500	—	28	3.3	7 to 200	16	—	180	2,200	22	1983	1,300
Texas													
Ford Geraldine	Ramsey-SS	2,680	83	40	1.4	64	23	60	23	900	26	1981	5,280
Kurten, Brazos Co.	Woodbine SS	8,300	230	38	0.38	1	12	—	—	—	50	1981	250
Rankin	SS	7,900	—	37	0.6	300	27	—	—	—	11	1981	80
North Cowden, Ector Co.	Grayburg-Dol	4,300	94	35	1.5	7	11	409	125	—	40	1973	37,000

Field	Formation												
Crossett, Crane Co.	Devonian-Ch	5,300	106	44	0.4	5	22	110	80	1,650	40	1972	1,120
South Gillock, Galveston Co.	Frio-SS	9,000	214	38	0.4	900	28	61	36	750	40	1972	5,900
Kelly-Snyder, Scurry Co.	Canyon-LS	6,700	132	42	0.4	19	8	213	139	1,600	51	1972	50,000
Mead-Strawn, Jones Co.	Strawn-SS	4,475	135	41	1.3	9	9	—	9	850	33	1964	3,900
Slaughter, Hockley Co.	San Andres-Dol	4,950	105	28	2.1	8	10	150	89	1,075	34	1976	87,000
Two Freds, Loving Co.	Delaware-SS	4,800	104	36	1.5	32	20	42	25	1,400	40	1974	4,400
Wasson, Yoakum Co.	San Andres-Dol	4,890	107	32	1	2	11	450	111	1,250	20	1972	63,500
Levelland, Hockley Co.	San Andres-Dol	4,750	102	30	2.3	3	11	—	80	1,050+	14	1973	—
South Welch		9,000	231	37	0.86	10 to 100	13	—	—	—	15		370
West Virginia													
Granny's Creek, Clay Co.	Big Injun-SS	2,000	75	45	—	5	18	40	34	1,000	10	1976	3,000
Griffithville, Lincoln Co.	Berea-SS	2,300	83	43	3.1	8	11	22	12	1,000	10	1976	10,000
Rock-Creek, Roane Co.	Big Injun-SS	2,000	73	40	1	20	22	20	35	1,000	10	1976	11,200
Wyoming													
N. Meadow Creek		3,500	106	40	1	20	21	—	—	920	—	1963	40
France													
Pecorade	Barremian-LS	7,500	195	30	0.6	1	10	—	—	—	5	1981	20
Hungary													
Budafa	Felső Lispe	3,230	154	42.5	1.16	22	22	—	—	—	5	1969	148
Budafa	Budafa	2,780	68	43	2.4	100	21	—	—	—	12	1972	865
Budafa	Kiscsehi	2,760	68	42.5	2.4	50	20	—	—	—	15	1974	222
Budafa	Zala Kerettye	2,952	147	42.5	0.8	70	22	—	—	—	13	1981	1,013
Lovászi	Lovászi Kelet	4,215	181	40.3	0.4	30	17	—	—	—	13	1975	346
Lovászi	Lovászi Nyugat	4,336	181	40.3	0.4	30	17	—	—	—	14	1977	568
Rumania													
Satchinez	Panonian-SS	4,260	176	48	0.7	10	25	—	26	—	—	1982	—
Calacea	Miocene-SS	3,600	140	48	0.8	1.3	25	—	72	—	—	1982	—

Sources: ''Annual Production Report,'' Oil and Gas J. (Apr. 5, 1982); Desch et al., ''Enhanced Oil Recovery by CO_2 Miscible Displacement in the Little Knife Field, Billings County, North Dakota,'' paper SPE/DOE 10696, Tulsa (1982); L. W. Holm and L. J. O'Brien, ''Carbon Dioxide Test at the Mead-Strawn Field,'' J. Pet. Tech. (Apr. 1971); A. V. Kane, ''Performance Review of a Large-Scale CO_2-WAG Enhanced Recovery Project, SACROC Unit-Kelly-Snyder Field,'' J. Pet. Tech. (Feb. 1979); G. E. Perry, ''Weeks Is and 'S' Sand Reservoir B Gravity Stable Miscible CO_2 Displacement, Iberia, Parish, Louisiana,'' paper SPE/DOE 10695, Tulsa (1982); S. B. Pontius and M. J. Tham, ''North Cross (Devonian) Unit CO_2 Flood-Review of Flood Performance and Numerical Simulation Model,'' J. Pet. Tech (Dec. 1978); Rowe et al., ''Slaughter Estate Unit Tertiary Pilot Performance,'' J. Pet. Tech. (Mar. 1982).

water and oil production by water reinjection. Injected gas density was cut with the addition of natural gas since CO_2 density (at reservoir pressure and temperature) was approximately equal to the in-place density of the S-Reservoir oil. After the first three years of injection, 31,851 STB of oil have been recovered with an estimated ultimate recovery of 65,000 STB.

6.4.2   **Immiscible Applications.** A number of immiscible CO_2 injection field tests and commercial developments have been presented in the literature. Table 6.6 lists nine of these projects. For more detailed information concerning individual tests, the reader is referred to the sources listed in the table. Two tests are worthy of additional note.

First, the Lick Creek Project in Arkansas is the largest commercial CO_2–heavy oil (17°) displacement test. The Meakin sandstone reservoir is a fault-trap structure with a formation depth of 2,550 ft, net thickness of 9 ft, permeability averages 1,200 md, porosity is 33%, and the connate water saturation is 32%. Oil viscosity at reservoir temperature and pressure is 160 cp, and the OOIP totals 23.3 MMSTB, with a projected primary recovery of 19% OOIP by pressure depletion and weak water drive (Kantar et al. 1983).

Injection began in February 1976 and implemented a WAG injection scheme. The operators have projected a total purchase of 8.5 BSCF of CO_2 to recover 3.09 MMSTB of incremental oil (2.75 MSCF/STB), almost doubling the recovery performance predicted by primary depletion alone. Half the total gas injected is from recycling operations, and a 15-year flood life is predicted.

Second, Turkey's Bati Raman Project will be split into cyclic stimulation and WAG drive. The field, discovered in 1961, contains low pressure 12 to 13°API gravity from the Garzan limestone at a depth of 4,300 ft. It is the largest oil field in Turkey with an estimated initial reserve of 1.85 BSTB of oil and a predicted primary depletion recovery factor of 1.5% OOIP (Kantar et al. 1983).

Applications of both steam or CO_2 appeared favorable. However, immiscible CO_2 injection appears more economically sound due to the nearby Dodan CO_2 gas field. The western test area, for cyclic stimulation, covers 1,200 acres with 32 wells. The central test area will encircle nine fully enclosed 62-acre five spots and will affect 990 acres. Injection operations are scheduled to begin in early 1984, with projected ultimate recovery ranging between 17% and 32% of the original oil in place.

These tests show a wide range of reservoir oils, formations, and operating conditions where CO_2 flooding has been introduced. The following list, then, summarizes and collates reservoirs applying CO_2 under similar conditions:

Light oil, moderate temperature (100 to 150°F),
irregular carbonates and sandstone reservoirs
Mead Strawn
Kelly-Snyder
Crossett
Two Freds
Wasson-Levelland

Table 6.6 Immiscible CO_2 Projects

Field	State	Operator	Test Size, Acres	Pay Zone	Depth, ft	Oil Gravity, degrees API
Huntington Beach	California	Aminoil	17	100 sandstone	2,700	16
Wilmington	California	Champlin	40.5	Tar sandstone	2,500	14
Bati Raman	Turkey	Turkish Petroleum	2,190	Mardin limestone	3,600	12
West Coyote	California	Union	5	—	3,300	21
Wainwright	Alberta	Husky	32	Sparky G sandstone	1,500	12
Ritchie	Arkansas	Phillips	220	Baker sandstone	2,600	16
Lick Creek	Arkansas	Phillips	1,640	Meakin sandstone	2,550	17
Grenade	France	Elf	—	Vraconien sandstone	7,700	11
Bradu	Rumania	—	—	Meotian sandstone	3,300	24

Sources: "Annual Production Report," *Oil and Gas J.* (Apr. 5, 1982); Kantar et al., "Heavy Oil Recovery by CO_2 Application from Bati Raman Field, Turkey," paper SPE 11475, Bahrain (1983); Khatib et al., "CO_2 Injection as an Immiscible Application for Enhanced Oil Recovery in Heavy Oil Reservoirs," paper SPE 9928, Bakersfield (1981); Patton et al., "Carbon Dioxide Well Stimulation: Part 2—Design and Aminoil's North Bolsa Strip Project," *J. Pet. Tech.* (Aug. 1982); T. B. Reid and H. J. Robinson, "Lick Creek Meakin Sand Unit Immiscible CO_2 Waterflood Project," *J. Pet. Tech.* (Mar. 1982).

High temperature (225°F), deep (>10,000 ft)
Little Creek
Weeks Island
Citronelle

Low temperature (<100°F)
Rock Creek
Tinsley
McCallum Unit
Maljamar

High gas saturation (30%)
Crossett

Gravity stable
Weeks Island
Bay St. Elaine
Paradis

Heavy oil
Wilmington (14°API)
Lick Creek (16°API)
East Coyote (22°API)
Bati Raman (12°API)

Low permeability (5 md)
Crossett
Levelland
Wasson
Kurten

Tertiary
Little Creek
Kelly-Snyder
Two Freds
Slaughter Estate
North Coles Levee
Fort Geraldine

Full scale
Kelly-Snyder
Crossett
Two Freds
Fort Geraldine
Lick Creek
Bati Raman

Note the wide variety of reservoir test properties: secondary and tertiary applications, light and heavy oils, sandstones and carbonates, hot (>225°F) and cold (<100°F) formations, flat and steeply dipping reservoirs. Also, different operating conditions have been employed, including continuous, slug, and WAG injection schemes with various slug sizes and CO_2/water injection ratios.

Given this broad spectrum of CO_2 experience, the reader is then encouraged to review particular applications of the CO_2 process that apply to his or her specific needs.

6.4.3   **Reservoir Response.** Table 6.7 summarizes the results of selected field tests where the information shown on the table was available by either direct measure or by simulation. The SACROC Main, Crossett, Two Freds, and Lick Creek projects were commercial-size floods, while the others are small-scale pilot tests of less than 100 acres.

CO_2 breakthroughs have occurred after injection of only 5 to 15% HCPV of total fluids. This is consistent with observations from past field tests of hydrocarbon miscible displacement. For a more detailed view of CO_2 production (percentage of the cumulative injected CO_2 that is produced as a function of total HCPV injected), see figure 6.6 (Reid and Robinson 1981; Stalkup 1983). This figure compares the CO_2 breakthrough and cumulative CO_2 production behavior of the SACROC, Willard-Wasson, Little Creek, and Lick Creek floods.

As stated, early CO_2 breakthrough was noted in all of these projects and was attributed to high CO_2 mobility, viscous fingering, gravity override, and/or reservoir heterogeneities. As the floods progressed, a large fraction of the injected slug was produced (30% to 70%). This would lead one to believe that significant recycling of injectant should be anticipated.

Table 6.7 Results From Selected Field Tests

Project	Slug Size (% HCPV)	Breakthrough (% HCPV)	Incremental Recovery (% OOIP)	Gross CO_2/Oil Ratio (MSCF/STB)	Net CO_2/Oil Ratio (MSCF/STB)
SACROC main flood process phases I and II	12 to 15	1.3 to 1.6	7[a]	6 to 7[a]	4.6[a]
SACROC tertiary pilot	10 to 18	5	3.5	12 to 20	5 to 14
Crossett	40	16	1.35 and increasing	7 to 9	7 to 9
Willard Wasson	20	10[a]	8 to 12	5 to 7[a]	3 to 4[a]
Slaughter Estate	26	10 to 15	15 and increasing (25[a])	10 and decreasing	—
Two Freds	25 and increasing	5	3 and increasing	26	—
Little Creek	160	15	18	24	13.5
Lick Creek	112 and increasing	—	3.3 and increasing (13[a])	13 to 18	7 to 10 (2.75[a])

[a]Calculated with flood simulators utilizing test data.

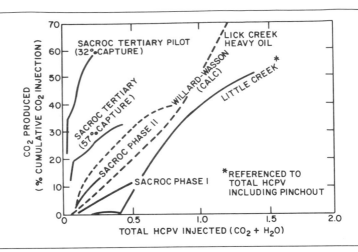

Figure 6.6 CO$_2$ production in field tests (Reid and Robinson 1981; Stalkup 1983).

Table 6.7 also shows the incremental recovery (percentage OOIP) for the few field tests from which information is available. Oil recoveries, both actual and projected, range from about 4% to 12% OOIP for slug sizes of 12% to 40% HCPV and are as high as 18% OOIP at Little Creek where a 160% HCPV slug was injected. Figure 6.7 shows these incremental production results for several field tests (Reid and Robinson 1981; Stalkup 1983).

The most successful project appears to be the Slaughter Estate Unit tertiary pilot where incremental recovery was reported to be 15% OOIP (and climbing) for an injected CO$_2$ slug of 26% HCPV. Low recovery (3% OOIP and rising) was expected for the heavy oil Lick Creek commerical test; however, ultimate incremental recovery is projected to be over 13% OOIP after 15 years of injection and recycling, while primary depletion produced only 19% OOIP.

Note that figure 6.7 also shows that the bulk of incremental oil production occurs with or after CO$_2$ breakthrough rather than being banked ahead of the injected slug. It appears from these results that early CO$_2$ breakthrough can be anticipated with significant reinjection of produced gas and an extended period of incremental oil recovery.

Last, table 6.7 shows that the ratio of gross CO$_2$ injected per STB of incremental oil recovered ranged from 5 to 26 (MSCF/STB) for moderate slug sizes of 12 to 40% HCPV. With the expected high recycling of produced gas comes a reduction in the net CO$_2$ injected per STB (3 to 14 MSCF/STB). This CO$_2$/oil ratio is similar to steam/oil and air/oil ratios for steam flooding and in situ combustion and is a prime indicator of economic success.

In summary, Stalkup (1983) has noted that one should expect to see a spectrum of behavior in CO$_2$ field tests. The incremental oil recovery and the CO$_2$/oil ratio will depend on factors such as volumetric sweep of the CO$_2$, swept zone residual oil saturations left after waterflooding and CO$_2$ flooding, the efficiency with which displaced oil is captured, the oil formation volume factor, and the CO$_2$ fvf factor. All of these vary from reservoir to reservoir. Incremental recovery and the CO$_2$/oil ratio also depend

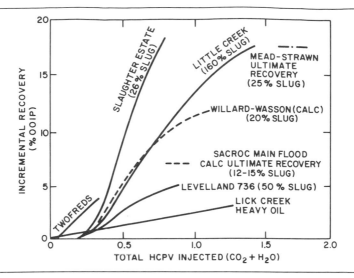

Figure 6.7 Incremental oil production in CO_2 field tests (Reid and Robinson 1981; Stalkup 1983).

on slug size, and for each project there will be an optimum slug size that results in optimum economic return.

One study by Bloomquist et al. (1981) discusses this optimum slug size selection procedure. Using a composition simulator, they projected oil recovery for various CO_2 slug sizes. They then input this information into an after tax economic package and selected the slug size that maximized rate of return. Their work, shown in figure 6.8, identifies oil price and CO_2 cost as major variables in determining the optimum slug size for a given reservoir.

6.5   **FINAL DESIGN.** Information from pilot test analysis and additional laboratory testing must now be incorporated in the final design process. This information is used to fine tune simulation efforts to narrow the range of expected recovery performance.

Plant design, economic re-evaluation, and development of the project are the last hurdles to commerical operation. A number of important factors will affect the economic viability of that design. First, of course, are the physical characteristics of the oil field, including reservoir rock and fluid properties, depth, size, state of development, and producing/operating history. The selection, design, and operation of the process will also be important. Project economics will be highly dependent on the expected future price of oil and the cost of the injectant.

In addition to the important physical and free market economic factors that must be considered, industry is now faced with a tax/regulatory situation that affects the eco-

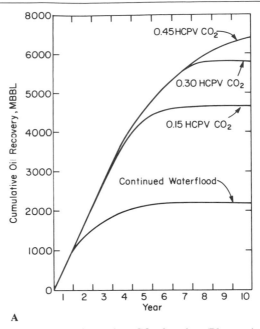

A

Figure 6.8A Projected oil recovery for various CO_2 slug sizes (Bloomquist et al. 1981).

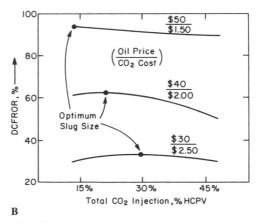

B

Figure 6.8B Discounted cash flow rate of return after federal income tax versus total CO_2 injection for standard windfall profits assumptions (Bloomquist et al. 1981).

nomic analysis of all potential projects. The windfall profit tax alone, and possible future changes to that tax, have significant cash flow implications.

In the final analysis, however, many questions must be answered during all phases of the CO_2 design process: data gathering, preliminary screening, laboratory testing, numerical simulation, and pilot testing with continuous economic re-evaluation.

This text pointed out the critical questions that must be asked and potential sources of information to resolve them. Any CO_2 design checklist must, from the onset, include

What CO_2 process could we employ, if any—heavy oil immiscible, multiple-contact miscible, or assumed first-contact miscible? secondary or tertiary? Check correlations to estimate miscibility pressure. Reaffirm with slim tube experiments. Are these pressures attainable in the field?

How does it compare with other EOR processes? For light oils, inert gas injection such as N_2 will require higher pressures, while chemical methods will require high front-end costs. With heavy oils, CO_2 may be competitive with steam in deep or thin pays.

How does it compare with waterflooding? incremental oil recovery?

What laboratory tests should be run to identify the displacement mechanisms? slim tubes, cores, swelling and viscosity information, single- and/or multiple-contact equilibrium?

How much oil is in place?

Can I estimate oil recovery without simulation? Hand-held methods offer rough estimates only.

Do we have a reliable CO_2 source? natural or manufactured?

What are the costs: CO_2, pipelines, transportation, drilling, recompletion, compressors, dehydrators, separators?

Should we run numerical simulators, and if so, what type? black oil immiscible, compositional, black oil miscible, or hybrid?

What type of injection scheme should be employed? Continuous (gravity stable), slug, WAG, simultaneous injection, huff-n-puff? Only pilot testing can verify that significant quantities of oil were mobilized.

Timing of recovery with respect to pattern size and type? Usually maintain in place pattern (5 to 40 acres), line drive (gravity stable), five spot, seven spot, nine spot.

If slug-type injection, determine the optimum slug size. Usually 12 to 40% HCPV chased by H_2O in horizontal floods or by N_2 or flue gas in gravity stable displacements.

If WAG-type injection, what is the proper water-gas injection ratio? The ratio is determined by saturations that yield a velocity of injected water slightly less than or equal to the velocity of CO_2 slug. Will injected water shield CO_2-oil contact?

If WAG-type injection, how many slugs of CO_2 should be injected and what size? Only total CO_2 injected appears to affect overall oil recovery. Changing the number of slugs or their size shows no apparent change in oil recovery. To start, one might suggest two to four equal-sized slugs and vary simulation parameters from there.

For gravity stable displacements, how does flood-front velocity compare to critical velocity? As in the Bay St. Elaine and Weeks Island projects, methane may be added to injected CO_2 to lower flood-front velocity.

Will gravity override be a problem in horizontal displacements? CO_2 and resident oil densities may be similar at reservoir temperature and pressure. However, if CO_2 and H_2O are injected together (WAG), gravity segregation may occur. Estimate the viscous/gravity force ratio for figure 5.13.

Is a preflush needed to increase reservoir pressure? Water injection has been used in projects like SACROC to raise reservoir pressure near the miscibility pressure prior to CO_2 injection.

Should we stage start up? Staging the number of patterns to go on line at one time will decrease the required field delivery rate of CO_2 but will also delay ultimately recovery.

Should produced gases be separated and dehydrated, with the CO_2 recycled? Not if impurities are in small quantities; otherwise, yes, sell natural gas and dispose of H_2S.

If a pilot test is run, what tracers should be used and which reservoir properties are critical? Organic and inorganic salts, alcohols, or radioactive isotopes are used to determine S_o, injection profile, and Kh, directional permeability and injectivity.

Can pilot test results and past production history be matched by the chosen reservoir simulator? A matched simulation effort is necessary to examine the large number of operating schemes possible. Coupled with an economic model, one can determine the type of injection scheme to employ, slug size, pattern size, pattern type, and so forth.

Are there unforeseen operating problems? Corrosion, asphaltenes, compressor and pipelines hydrates, H_2S detection and handling, dehydration, and separation.

What capacity of produced fluids must be handled?

Can we sell the oil and at what price?

Is there political instability and what will future taxes be?

What is the cost of capital?

Will the target profitability be reached?

REFERENCES

Adams, G. H., and Rowe, H. G.: "Slaughter Estate Unit CO_2 Pilot-Surface and Downhole Equipment Construction and Operation in the Presence of H_2S Gas," paper SPE 8830 presented at the SPE/DOE 1st Joint Symposium on Enhanced Oil Recovery, Tulsa, April 20–23, 1980.

"Annual Production Report," *Oil and Gas J.* (April 5, 1982) 139–159.

Bloomquist, C. W., Fuller, K. L., and Moranville, M. B.: "Miscible Gas EOR Economics and Effects of the Windfall Profits Tax," paper SPE 10274 presented at the SPE 56th Annual Meeting, San Antonio, Oct. 5–7, 1981.

Christian, L. D., Shirer, J. A., Kimble, E. J., and Blackwell, R. J.: "Planning a Tertiary Oil Recovery Project for the Jay-Little Escambia Creek Fields Unit," paper SPE/DOE 9805 presented at the SPE/DOE 2nd Joint Symposium on Enhanced Oil Recovery, Tulsa, April 5–8, 1981.

Desch, J. B., Larsen, W. K., Lindsay, R. F., and Nettle, R. L.: "Enhanced Oil Recovery by CO_2 Miscible Displacement in the Little Knife Field, Billings County, North Dakota," paper SPE/DOE 10696 presented at the SPE/DOE 3rd Joint Symposium on Enhanced Oil Recovery, Tulsa, April 4–7, 1982.

Drennon, M. D., Kelm, C. H., and Whitington, H. M.: "A Method for Appraising the Feasibility of Field-Scale CO_2 Miscible Flooding," paper SPE 9323 presented at the SPE 55th Annual Technical Conference, Dallas, Sept. 21–24, 1980.

Graham, B. D., Bowen, J. F., Duane, N. C., and Warden, G. D.: "Design and Implementation of a Levelland Unit CO_2 Tertiary Project," paper SPE 8831 presented at the SPE/DOE 1st Joint Symposium on Enhanced Oil Recovery, Tulsa, April 20–23, 1980.

Holm, L. W., and O'Brien, L. J.: "Carbon Dioxide Test at the Mead-Strawn Field," *J. Pet. Tech.* (April 1971) 431–442.

Intercomp Consortium Report: "Feasibility of CO_2 Enhanced Oil Recovery in the Rocky Mountain Region," Intercomp, Denver (1980).

Kane, A. V.: "Performance Review of a Large-Scale CO_2-WAG Enhanced Recovery Project, SACROC Unit-Kelly-Snyder Field," *J. Pet. Tech.* (Feb. 1979) 217–231.

Kantar, K., Karaoguz, D., Issever, K., and Vrana, L.: "Heavy Oil Recovery by CO_2 Application from Bati Raman Field, Turkey," paper SPE 11475 presented at the Middle East Oil Technical Conference, Bahrain, March 14–17, 1983.

Khatib, A. K., Earlougher, R. C., and Kantar, K.: "CO_2 Injection as an Immiscible Application for Enhanced Oil Recovery in Heavy Oil Reservoirs," paper SPE 9928 presented at the SPE California Regional Meeting, Bakersfield, March 25–26, 1981.

Koval, E. J.: "A Method for Predicting the Performance of Unstable Miscible Displacement in Heterogeneous Media," Soc. Pet. Eng. J. (June 1963) 145–154.

Kuuskraa, V. A., Hammershaimb, E. C., and Wicks, D. E.: "EOR Major Equipment and Its Projected Demand," Oil and Gas Petrochem. Equip. (Sept. 1981) 28–29, 48–49.

Lewin and Associates, Inc.: "Economics of Enhanced Oil Recovery," report DOE/ET/78-C-01-2628, Final Report for U.S. DOE (March 1981).

Mungan, N.: "Enhanced Oil Recovery Using Water as a Driving Fluid," World Oil (Sept. 1981) 155–167.

National Petroleum Council: Enhanced Oil Recovery, NPC, Washington, D.C. (1976).

Orr, F. M., Silva, M. K., Lien, C. L., and Pelletier, M. T.: "Laboratory Experiments to Evaluate Field Prospects for CO_2 Flooding," J. Pet. Tech. (April 1982) 888–903.

Patton, J. T., Sigmund, P., Evans, B., Ghose, S., and Weinbrandt, D.: "Carbon Dioxide Well Stimulation: Part 2—Design and Aminoil's North Bolsa Strip Project," J. Pet. Tech. (Aug. 1982) 1805–1810.

Peng, D. Y., and Robinson, D. B.: "A New Two-Constant Equation-of-State," Ind. and Eng. Chem. Fund. (1976) 15, 59–64.

Perry, G. E.: "Weeks Island 'S' Sand Reservoir B Gravity Stable Miscible CO_2 Displacement, Iberia, Parish, Louisiana," paper SPE/DOE 10695 presented at the SPE/DOE 3rd Joint Symposium on Enhanced Oil Recovery, Tulsa, April 4–7, 1982.

Pontious, S. B., and Tham, M. J.: "North Cross (Devonian) Unit CO_2 Flood-Review of Flood Performance and Numerical Simulation Model," J. Pet. Tech. (Dec. 1978) 1706–1715.

Pullman-Kellogg, Inc.: "Sources and Delivery of Carbon Dioxide for Enhanced Oil Recovery," DOE Report FE-2515-24, Washington, D.C. (1978).

Redlich, O., and Kwong, J. N. S.: Chem. Rev. (1949) 44, 233.

Reid, T. B., and Robinson, H. J.: "Lick Creek Meakin Sand Unit Immiscible CO_2 Waterflood Project," J. Pet. Tech. (Sept. 1981) 1723–1729.

Rowe, H. G., York, S. D., and Ader, J. C.: "Slaughter Estate Unit Tertiary Pilot Performance," J. Pet. Tech. (March 1982) 613–620.

Soave, G., "Equilibrium Constants from a Modified Redlich-Kwong Equation of State," Chem. Eng. Sci. (1972) 27, 1197–1203.

Stalkup, F. I., Jr.: Miscible Displacement, Monograph Series, SPE, Dallas (1983) 8, 137–158.

Taber, J. J., and Martin, F. D.: "Technical Screening Guides for the Enhanced Recovery of Oil," paper SPE 12069 presented at the SPE 58th Annual Technical Conference and Exhibition, San Francisco, Oct. 5–8, 1983.

Todd, M. R., and Longstaff, W. J.: "The Development, Testing and Application of a Numerical Simulator for Predicting Miscible Flood Performance," J. Pet. Tech. (July 1972) 874.

Wilson, G. M.: "A Modified Redlich-Kwong Equation of State, Applications to General Physical Data Calculations," paper No. 15C presented at the American Institute of Chemical Engineers 65th National Meeting, Cleveland, May 4–7, 1969.

Yarborough, L.: "Application of a Generalized Equation of State to Petroleum Reservoir Fluids," Equation of State in Engineering and Research, K. C. Chao and R. L. Robinson, Jr. (eds.), Advances in Chemistry Series, Am. Chem. Soc. (1978) 182, 386–439.

Zudkevitch, D., and Jaffe, J.: "Correlation and Prediction of Vapor-Liquid Equilibria with the Redlich-Kwong Equation of State," AIChE J. (1970) 16, 112.

INDEX